Online Storage Systems and Transportation Problems with Applications

Applied Optimization
Volume 91

Series Editors:

Panos M. Pardalos
University of Florida, U.S.A.

Donald W. Hearn
University of Florida, U.S.A.

Online Storage Systems and Transportation Problems with Applications

Optimization Models and
Mathematical Solutions

by

Julia Kallrath
ITWM, Germany

 Springer

Library of Congress Cataloging-in-Publication Data

A C.I.P. Catalogue record for this book is available
from the Library of Congress.

ISBN 1-4020-2971-3   e-ISBN 0-387-23485-3   Printed on acid-free paper.

Printed in the United States of America.

9 8 7 6 5 4 3 2 1   SPIN 11161943

springeronline.com

to my parents

Contents

Preface

This book covers the analysis and development of online algorithms involving exact optimization and heuristic techniques, and their application to solve two real life problems.

The first problem is concerned with a complex technical system: a special carousel based high-speed storage system - Rotastore. It is shown that this logistic problem leads to an NP-hard *Batch PreSorting Problem* (BPSP) which is not easy to solve optimally in offline situations. We consider a polynomial case and develope an exact algorithm for offline situations. Competitive analysis showed that the proposed online algorithm is 3/2-competitive. Online algorithms with lookahead improve the online solutions in particular cases. If the capacity constraint on additional storage is neglected the problem has a totally unimodular polyhedron.

The second problem originates in the health sector and leads to a vehicle routing problem. We demonstrate that reasonable solutions for the offline case covering a whole day with a few hundred orders can be constructed with a heuristic approach, as well as by simulated annealing. Optimal solutions for typical online instances are computed by an efficient column enumeration approach leading to a set partitioning problem and a set of routing-scheduling subproblems. The latter are solved exactly with a branch-and-bound method which prunes nodes if they are value-dominated by previous found solutions or if they are infeasible with respect to the capacity or temporal constraints. Our branch-and-bound method is suitable to solve any kind of sequencing-scheduling problem involving accumulative objective functions and constraints, which can be evaluated sequentially. The column enumeration approach developed to solve this hospital problem is of general nature and thus can be embedded into any decision-support system involving assigning, sequencing and scheduling.

The book is aimed at practioners and scientists in operation research especially those interested in online optimization. The target audience are readers interested in fast solutions of batch presorting and vehicle routing problems or software companies producing decision support systems. Students and graduates in mathematics, physics, operations research, and businesses with interest in modeling and solving real optimization problems will also benefit from this book and can experience how online optimization enters into real world problems.

Structure of this Book

This book is organized as follows. Chapter 2 addresses the BPSP, where a formal definition of the BPSP is introduced (Section 2.1) and several modeling approaches are proposed (see Section 2.2). Complexity issues of some formulations are investigated in Section 2.3 and Section 2.4. For one polynomial case of the BPSP several algorithms are presented and compared in Section 2.5. In Chapter 3 we consider a concrete application of the BPSP - carousel based storage system Rotastore. In Section 3.1 we describe the system performance, and in Section 3.2 the numerical results of the experiments are presented.

Chapter 4 focuses on the *Vehicle Routing problem with Pickup and Delivery and Time Windows* (VRPPDTW), adapted for hospital transportation problems. After introducing some notations (Subsection 4.2.1), we suggest several approaches we have developed to solve this problem, including a MILP formulation (Subsection 4.3.1), a branch-and-bound approach (Subsection 4.3.2), a column enumeration approach (Subsection 4.3.3), and heuristic methods (Section 4.4). In Chapter 5 we describe a problem related to a hospital project with the University Hospital in Homburg. Detailed numerical results for our solution approaches related to the VRPPDTW are collected in Section 5.2.

Conventions and Abbreviations

The following table contains in alphabetic order the abbreviations used in this book.

Abbreviation	Meaning
B&B	Branch-and-Bound
B&C	Branch-and-Cut
BPSP	Batch PreSorting Problem
CEA	column enumeration approach
IP	Integer Programming
LP	Linear Programming
MCP	Mixed Complementarity Problem
MILP	Mixed Integer Linear Programming
MINLP	Mixed Integer Nonlinear Programming
RH	reassignment heuristic
SA	simulated annealing
SAT	satisfiability problem
SH	sequencing heuristic
s.t.	subject to
TS	tabu search
VNS	variable neighborhood search
VRP	Vehicle Routing Problem
VRPPDTW	VRP with Pickup and Delivery and Time Windows
w.r.t.	with respect to

Acknowledgements

First of all I want to thank Prof. Dr. Stefan Nickel (ITWM Kaiserslautern and Universität Saarbrücken), the head of our department, who was leading the two real world projects; without him the book would not have appeared. Prof. Dr. Christodoulos Floudas (Princeton University) and Prof. Dr. Linus Schrage (University of Chicago) - for their interest in this work and encouraging comments on the vehicle routing problem. Prof. Dr. Susanne Albers (Freiburg Universität) - for the useful feedback on the topic of online optimization and the competitive analysis for the batch pre-sorting problem. Prof. Dr. Alexander Lavrov (ITWM Kaiserslautern and NTUU Kiev) - for his constant support and help during all my time at the ITWM in Kaiserslautern. Dr. Teresa Melo (ITWM Kaiserslautern)- for fruitful discussions about the vehicle routing problem and for the proof reading. Martin Müller (Siemens AG, München) - for constructive talks and discussions about the batch pre-sorting problem. Prof. Dr. Robert E. Wilson (University of Florida), Steffen Rebennack (Universität Heidelberg) and Dr. Anna Schreieck (BASF, Ludwigshafen) - for careful reading parts of this book. And last but not least, my husband Josef Kallrath - for his positive spirit encouraging and supporting me never to give up.

JULIA KALLRATH

Chapter 1

INTRODUCTION

What do a logistics manager responsible for an inventory storage system and a vehicle fleet dispatcher in a hospital campus have in common? They both have to consider new objects arriving at short notice and to decide on what to do with them, how to assign them to given resources or how to modify previously made decisions. This means they both need to make decisions based on data suffering from incomplete knowledge about near future events. Online optimization is a discipline in mathematical optimization and operations research which provides the mathematical framework and algorithms for dealing appropriately with such situations.

1.1.   Optimization Everywhere

The need for applying optimization arises in many areas: finance, space industry, biosystems, textile industry, mineral oil, process and metal industry, and airlines to name a few. Mathematical programming is a very natural and powerful way to solve problems appearing in these areas. In particular, see [12], [18], [23], [37] and [83] for application examples. One might argue that low structure systems can probably be handled well without optimization. However, for the analysis and development of real life complex systems (that have many degrees of freedom, underlying numerous restrictions *etc.*) the application of optimization techniques is unavoidable. It would not be an exaggeration even to say that any decision problem is an optimization problem. Despite their diversity real world optimization problems often share many common features, *e.g.*, they have similar mathematical kernels such as flow, assignment or knapsack structures.

One further common feature of many real life decision problems is the online nature aspect, *i.e.*, decision making is based on partial, insufficient information or without any knowledge of the future. One approach (not treated in this book) to solve problems with only partial or insufficient information is optimization under uncertainty (*cf.* [45], [50], or [88]), and especially, stochastic programming (*cf.* [14], [53], [77], or [78]). In that case, the problem is still solved as an offline problem.

However, it is not always appropriate to solve a problem offline. If we cannot make any assumptions on future data, only the currently available data can be used. In such situations online optimization is recommended. We can list a number of problems that were originally formulated as offline problems but which in many practical applications are used in their online versions: the bin packing problem, the list update problem, the k-server problem, the vehicle routing problem, and the pickup and delivery problem to name a few.

Special optimization techniques for online applications exploit the online nature of the decision process. Usually, a sequence of online optimization problems is solved when advancing in time and more data become available. Therefore, online optimization can be much faster than offline optimization (which uses the complete input data). To estimate the quality of a sequence of solutions obtained by online optimization one can only compare it with the overall solution produced by an offline algorithm afterwards. A powerful technique to estimate the performance of online algorithms is the competitive analysis (*cf.* [11]). A good survey on online optimization and competitive analysis can be found in [4], [11], [30]. Online optimization and competitive analysis are based on generic principles and can be beneficial in completely different areas such as the storage system and transportation problem considered in this book.

At first we consider an example of a complex technical system, namely a special carousel based high-speed storage system - *Rotastore* [73], which not only allows storing ([56], [57]) but also performs sorting ([49], [70]). Sorting actions and assignment to storing locations are fulfilled in real time, but the information horizon may be rather narrow. The quality of the corresponding decisions strongly influences the performance of the system in general; thus the need to improve the quality of the decisions. Due to the limited information horizon online optimization is a promising approach to solve these problems.

In our second case study, the conditions for the decision making process in hospital transportation are similar: the orders often are not known in advance, the transportation network may be changed dynamically. The efficiency of order assignment and scheduling of the transport system can influence the operation of the whole hospital. That assumes, in this case, not only economical aspects, but, at first of all, human health and life issues.

As will be shown in this book, the mathematical base for the first problem is the *Batch PreSorting Problem* (BPSP), for the second one we naturally can use an online variant of the *Vehicle Routing problem with Pickup and Delivery and Time Windows* (VRPPDTW). The efficient application of the corresponding solution methods allows to improve the performance of both systems compared to the current real life situation.

Chapter 2

BATCH PRESORTING PROBLEMS. I MODELS AND SOLUTION APPROACHES

This chapter is organized as follows: at first, we describe the problem and give a short classification. In Section 2.2 different formulations of the BPSP are presented. In Subsection 2.2.2 we consider an optimization version of BPSP$_1$. In Subsection 2.2.3 we formulate BPSP$_2$ and BPSP$_3$ as decision problems and additionally introduce optimization models. The complexity status of BPSP$_2$ is investigated in Section 2.3, and in Section 2.4 we show that there is a polynomial version of the BPSP. Also we consider a special subcase of a BPSP with $N^L = 2$ in offline and online situations and present corresponding algorithms in Section 2.5. Finally, in Section 2.6, some results derived for BPSPs with $N^L = 2$ are adapted to general BPSP.

2.1. Problem Description and Classification

We consider the problem of finding a finite sequence of objects of different types, that guarantees an optimal assignment of objects to given physical storage layers with a pre-sorting facility of limited capacity. This problem will be called the *Batch PreSorting Problem* (BPSP), because the objects have to be sorted within one batch before they are assigned to the layers. After sorting, the object with number i will be assigned to layer i. For a more transparent presentation we speak of colors instead of types and thus consider all objects of type k as having the same color k. We present three types of BPSP with different objective functions. The objective function, z, of $BPSP_1$ minimizes the total number of layers not yet occupied by objects of a certain color k; such objects can be considered as occupying an empty layer (empty *w.r.t.* to k) at zero cost. Once each layer has an object of a given color, the cost does not change with further additions of that color. If the forgoing is true for all colors, z gives the number of all objects to be distributed minus those already assigned to the layers. In $BPSP_2$ the objective is to minimize the maximum number of objects of the same color on the same layer. Finally, $BPSP_3$ aims to minimize the sum of the maximum number of objects of the same color over all layers.

We use the following example to illustrate the problem:

EXAMPLE 2.1 *Suppose, there are six objects of two different colors in the input sequence (see Fig. 2.1.1) and three layers.*

Objects can be sorted within one batch, i.e., the objects 1, 2, 3 can be sorted, then they are assigned to the layers. After this the objects 4,5 and 6 can be sorted and assigned to the layers. Fig. 2.1.1 displays the content of the layers without pre-sorting. For this assignment the objective function value of $BPSP_1$ is 2, because the objects of the first batch occupy the layers at zero cost (layers were empty); the objects 4 and 5 occupy the layers 1 and 2, respectively, each with cost one, and object 6 occupies layer 3 at zero cost. The objective function value of $BPSP_2$ is 2, because the maximal number of objects of any color on all layers is 2. Finally, the objective function value of $BPSP_3$ is 4, because the maximal number of objects of the colors 1 and 2 over all layers is 2 for both colors. Clearly, this assignment is not optimal w.r.t. none of the three objective functions. The optimal objective function values for $BPSP_1$, $BPSP_2$, and $BPSP_3$ are 0, 1, and 2, respectively (see Fig. 2.1.2).

2.2.   Formulation of the Batch Presorting Problem

At first we introduce some notations used in this chapter:

- N^O is the number of objects of different colors in a given sequence. These objects are indexed by i or j (for simplicity the positions of the objects are identified by their index values). S_k is the set of objects of color k, and $i \in S_k$ means that the object at position i in the sequence (also called "i^{th} object" or "object i" for short) has color k;

- N^K is the number of colors;

- N^L is the number of layers;

- N^S is the capacity of the pre-sorting facility.

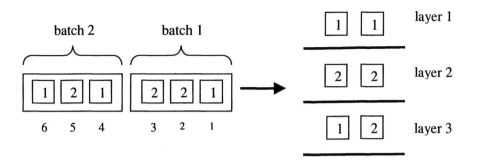

Figure 2.1.1. The input sequence and the content of the storage layers without pre-sorting. On the left part of the figure, the numbers 1, 2, ..., 6 refer to the objects while the numbers 1 and 2 in the squares denote the colors.

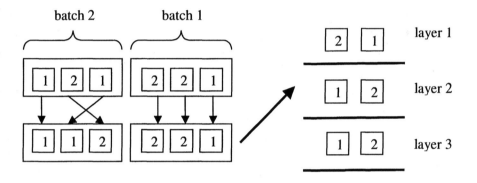

Figure 2.1.2. Optimal permutation and content of the storage layers after the assignment

2.2.1 Feasible Permutations

Before we talk about feasible permutations, recall the definition of permutation:

DEFINITION 2.2 *A permutation δ on a set of N^O objects is a one-to-one mapping of set $\{1,\ldots,N^O\}$ onto itself, i.e., $\delta : \{1,\ldots,N^O\} \longrightarrow \{1,\ldots,N^O\}$. Thus $\delta(i) = j$ if the object originally positioned at i, is placed onto position j.*

In other words, if δ is a permutation, $\delta(i)$ denotes the position of object i in the output sequence. In our case, only a subset of all possible permutations can be performed using the pre-sorting facility.[1]

---

[1] For the concrete technical functionality see Chapter 3.

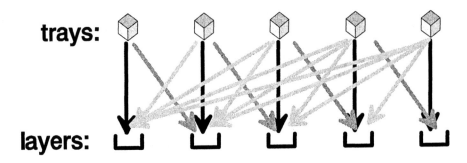

Figure 2.2.3. The set of all possible permutations for $N^S = 1$

THEOREM 2.3 *Let N^S be the capacity of the pre-sorting facility. A permutation δ is realizable, if and only if for each object i, $\delta(i) \geq i - N^S$. If $N^O \leq N^S$ then there exist $N^O!$ realizable permutations, otherwise there will be $N^S!(N^S + 1)^{N^O - N^S}$.*
Proof. *(see, for instance, [39])* ■

Fig. 2.2.3 illustrates the result of Theorem 2.3. Notice, that if $N^O \leq N^S$ then there exist $N^O!$ realizable permutations and $N^S!(N^S + 1)^{N^O - N^S}$ otherwise. In this work the terms *realizable* and *feasible* permutations are equivalent. Now we formally introduce the notion of a feasible permutation.

DEFINITION 2.4 *A permutation δ is feasible if for any $i = 1, ..., N^O$*

$$\delta(i) \geq i - N^S \qquad (2.2.1)$$

is fulfilled.

2.2.2   Mathematical Formulation of BPSP$_1$

As was defined above, only the permutations with $\delta(i) \geq i - N^S$ are feasible. Note that only the objects at the permuted positions $\delta(i) = j$ will be placed onto layer l, where $l \equiv j \bmod N^L$, i.e., l is a function of j. For example, if $N^O = 5$, $N^L = 2$, then objects with positions $j = 1, 3, 5$ will be placed onto layer $l = 1$, those with positions $j = 2, 4$ onto layer $l = 2$.

In addition we introduce the following notations:

$$\delta_{ij} = \begin{cases} 1, & \text{if } \delta(i) = j \\ 0, & \text{otherwise} \end{cases} ,$$

and

$$C'_{ij} = \begin{cases} 1, & \text{if layer } l, l \equiv j \bmod N^L, \text{ has already an object of the same} \\ & \text{color as object } i \\ 0, & \text{otherwise} \end{cases}$$

(2.2.2)

The optimal permutation can be constructed from the solution of the following linear program [39]:

$$\min \sum_{i=1}^{N^O} \sum_{j=1}^{N^O} C'_{ij} \delta_{ij} \quad , \tag{2.2.3}$$

$$\sum_{j=1}^{N^O} \delta_{ij} = 1, \quad 1 \le i \le N^O \quad , \tag{2.2.4}$$

$$\sum_{i=1}^{N^O} \delta_{ij} = 1, \quad 1 \le j \le N^O \quad , \tag{2.2.5}$$

$$\delta_{ij} = 0, \quad \forall i, j : j < i - N^S \quad , \tag{2.2.6}$$

$$\delta_{ij} \in \{0, 1\}, \quad 1 \le i, j \le N^O \quad . \tag{2.2.7}$$

We can interpret the coefficient C'_{ij} as the cost of placing object i onto position j (which uniquely identifies layer l). As (2.2.3) minimizes the total placing cost, it minimizes hence the total number of layers not yet occupied by objects of a certain color k. Such objects can populate an empty layer (empty *w.r.t.* to k) at zero cost. Infeasible permutations are excluded (depending on N^S), a priori by (2.2.6). Obviously, (2.2.6) corresponds to (2.2.1).

It is well known that this kind of integer program is totally unimodular (*cf.* [61]) and, thus, may be solved efficiently by some versions of the Simplex algorithm. Many special matching algorithms solve the problem in polynomial time (*cf.*, [72]). In practical applications (see Chapter 3), the performance very often depends on the number of attempts needed to output completely a set of orders (an order is a set of objects of different types). An attempt is considered successful if there exists at least one object of a given color on each layer (*i.e.*, belonging to the requested order). Therefore, for a given set of orders, the number of attempts needed for complete output is the maximum number of objects in these orders found on a single layer.

Consider, for instance, the following example: $N^O = 8$, $N^K = 2$ (*e.g.*, blue and yellow), $N^L = 2$, $C'_{ij} = 1$ for all i, j (*i.e.*, one blue and one yellow object already exist on each layer). Let the first four objects be

blue and the others yellow. Suppose, BPSP$_1$ has two optimal solutions
with objective function value 8:

1 Three blue objects are assigned to the first layer and one to the second; one yellow object to the first layer and three to the second.

2 Two objects of each color are assigned to both layers.

The numbers of attempts for complete output are $4 + 4 = 8$ in the first
case and $3 + 3 = 6$ in the second (see Fig. 2.2.4). In terms of suffi-

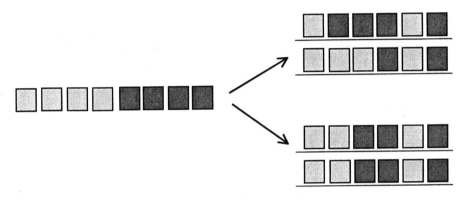

Figure 2.2.4. The example of two different assignments of objects to the storage
layers.

ciency the second solution is preferable, because it needs fewer attempts
for complete output. For practical applications we want to produce a
solution with minimal number of attempts. Since BPSP$_1$ does not necessarily do so, we developed the following problem formulations.

Note that the formulation above does not contain the index k, because
the information about the color of objects is hidden in the coefficients C'_{ij}.
More precise, $C'_{ij} = C'_{ij}(k)$. Example 2.5 illustrates how the coefficients
C'_{ij} are constructed.

2.2.3   Mathematical Formulation of BPSP$_2$ and BPSP$_3$

In this section we formulate BPSP$_2$ and BPSP$_3$ as decision problems.
Most of the notations used in the previous section will be kept. Analogous to the notation C'_{ij} from Section 2.2.2 we use the notation C_{kl} - the
number of objects of color k already present on layer l. Additionally we
define:

■ an integer bound B;

- constants

$$S_{ik} = \begin{cases} 1, & \text{if } i \in \mathcal{S}_k \\ 0, & \text{otherwise} \end{cases} , \qquad (2.2.8)$$

$$M_{jl} = \begin{cases} 1, & \text{if } j \equiv l \mod N^L \\ 0, & \text{otherwise.} \end{cases} \qquad (2.2.9)$$

This allows us to define

$$D_{ikjl} := S_{ik} M_{jl} \quad . \qquad (2.2.10)$$

Now we can formulate the following decision problems:

D-BPSP$_2$: Is there a feasible permutation δ such that the maximal cost

$$\max_{k=1,\ldots,N^K,\, l=1,\ldots,N^L} \left(C_{kl} + \sum_{i=1}^{N^O} \sum_{j=1}^{N^O} D_{ikjl} \delta_{ij} \right) \qquad (2.2.11)$$

does not exceed B?

D-BPSP$_3$: Is there a feasible permutation δ such that the total cost

$$\sum_{k=1}^{N^K} \max_{l=1,\ldots,N^L} \left(C_{kl} + \sum_{i=1}^{N^O} \sum_{j=1}^{N^O} D_{ikjl} \delta_{ij} \right) \qquad (2.2.12)$$

does not exceed B?

Remark: The term $D_{ikjl}\delta_{ij}$ takes the value 1 if an additional object i of color k is placed onto layer l by permutation δ. As C_{kl} denotes the number of objects of color k already present on that layer, the cost $C_{kl} + \sum_{i=1}^{N^O} \sum_{j=1}^{N^O} D_{ikjl}\delta_{ij}$ yields the number of objects after the permuted objects have all been placed in the layers. In other words, D-BPSP$_2$ is the problem of finding a permutation of objects such that the maximal number of objects of the same color on any layer is less than or equal to B for all colors. Thus, for practical applications, the total cost term of D-BPSP$_2$ can be interpreted as a worst-case estimation of the performance and the total cost of D-BPSP$_3$, analogously, represents the average performance over all colors.

2.2.3.1 An Optimization Version of BPSP$_2$

Since the objective is to minimize the maximal cost (2.2.11), we now formulate the decision problem as an optimization problem:

$$\min \, \max_{k=1,\ldots,N^K,\, l=1,\ldots,N^L} \left(C_{kl} + \sum_{i=1}^{N^O} \sum_{j=1}^{N^O} D_{ikjl} \delta_{ij} \right) \quad .$$

There appears to be no easy way to solve the above problem efficiently. Therefore we transform it to an integer linear programming formulation. The *minimax* objective function is replaced by an equivalent linear formulation. For this aim new variables are introduced: u_k — the maximal number of objects of color k on any layer, *i.e.*,

$$u_k = \max_{l=1,\dots,N^L} \left(C_{kl} + \sum_{i=1}^{N^O} \sum_{j=1}^{N^O} D_{ikjl}\delta_{ij} \right) \qquad (2.2.13)$$

and

$$y = \max_{k=1,\dots,N^K} \{u_k\} \quad .$$

That allows us to define the transformed objective function

$$\min y \quad , \qquad (2.2.14)$$

subject to additional constraints

$$u_k \leq y, \quad 1 \leq k \leq N^K \quad , \qquad (2.2.15)$$

$$C_{kl} + \sum_{i=1}^{N^O} \sum_{j=1}^{N^O} D_{ikjl}\delta_{ij} \leq u_k, \qquad \begin{matrix} 1 \leq k \leq N^K \\ 1 \leq l \leq N^L \end{matrix} \quad , \qquad (2.2.16)$$

$$\sum_{j=1}^{N^O} \delta_{ij} = 1, \quad 1 \leq i \leq N^O \quad , \qquad (2.2.17)$$

$$\sum_{i=1}^{N^O} \delta_{ij} = 1, \quad 1 \leq j \leq N^O \quad , \qquad (2.2.18)$$

$$\delta_{ij} = 0, \quad \forall i,j : j < i - N^S \quad , \qquad (2.2.19)$$

$$\delta_{ij} \in \{0,1\}, \quad 1 \leq i,j \leq N^O \quad . \qquad (2.2.20)$$

Let us make some remarks related to the above constraints:

Note that (2.2.14) and (2.2.15) imply the identity $y = \max_{k=1,\dots,N^K} \{u_k\}$. The inequalities (2.2.16) express that the number of objects of color k already on layer l plus a number of objects of color k assigned to this layer cannot be greater than the maximal number u_k of objects of color k on any layer. The assignment constraints (2.2.17)-(2.2.18) determine the permutations of the objects, *i.e.*, each object can take only one position in a new ordering and each new position can be filled only with one object. Depending on N^S, certain permutations can be excluded a priori by (2.2.19).

In some situations, when the difference between the number of objects of different orders is very large, it may not be advisable to minimize just the maximum number of objects of this set of orders found on a single layer. Instead, it is more efficient to minimize the total amount of output cycles. We treat this approach in the optimization problem BPSP$_3$ introduced below.

2.2.3.2   An Optimization Version of BPSP$_3$

The objective function corresponding to (2.2.12) is:

$$\min \sum_{k=1}^{N^K} \max_{l=1,\ldots,N^L} \left(C_{kl} + \sum_{i=1}^{N^O} \sum_{j=1}^{N^O} D_{ikjl} \delta_{ij} \right) \ .$$

Using the new variables u_k (2.2.13), we get:

$$\min \sum_{k=1}^{N^K} u_k \tag{2.2.21}$$

subject to (2.2.16)-(2.2.20).

Thus, it can be seen that BPSP$_2$ and BPSP$_3$ contain BPSP$_1$ as a kernel. Unfortunately, the polyhedrons of BPSP$_2$ and BPSP$_3$ are not integral and, hence, the complexity issues of these problems are very important. The following section addresses the complexity of BPSP$_2$.

EXAMPLE 2.5 *In this example we illustrate in detail the optimization models BPSP$_1$, BPSP$_2$, and BPSP$_3$ with possible optimal solutions using the following set of input data:*

The number of objects, N^O, is 6, the number of colors, N^K, is 3. These objects are grouped together in the sets: $S_1 = \{1\}$, $S_2 = \{2,3,4\}$, $S_3 = \{5,6\}$. The number of layers, N^L, is 3, and the capacity of the pre-sorting facility is $N^S = 2$. Coefficients reflecting the content of the layers have the following values: $C_{11} = 1$, $C_{21} = 1$, $C_{22} = 2$, $C_{33} = 1$, all other $C_{kl} = 0$. Fig. 2.2.5 shows the input sequence and the content of the layers before the assignment.

BPSP$_1$: For this model the construction of coefficients C_{kl} is simplified, so we do not need to calculate the number of objects of color k on layer l, but only need to indicate whether layer l has an object of color k (see 2.2.2). Regarding this definition, only the coefficients $C'_{11}, C'_{21}, C'_{22}, C'_{31}, C'_{32}, C'_{41}, C'_{42}, C'_{53}$, and C'_{63} have value 1.

$$f_1 = \min(\delta_{11} + \delta_{21} + \delta_{22} + \delta_{31} + \delta_{32} + \delta_{41} + \delta_{42} + \delta_{53} + \delta_{63}) \tag{2.2.22}$$

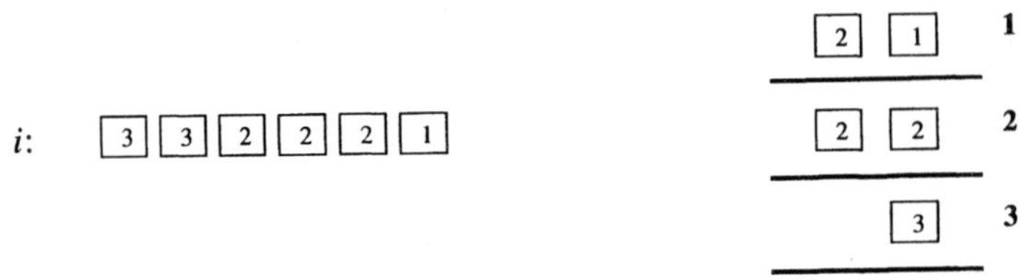

Figure 2.2.5. The input sequence and the content of the layers before the assignment

subject to

$$\sum_{j=1}^{6} \delta_{ij} = 1, \quad 1 \le i \le 6 \quad , \tag{2.2.22}$$

$$\sum_{i=1}^{6} \delta_{ij} = 1, \quad 1 \le j \le 6 \quad , $$

$$\delta_{41}, \delta_{51}, \delta_{52}, \delta_{61}, \delta_{62}, \delta_{63} = 0 \quad , \tag{2.2.24}$$

$$\delta_{ij} \in \{0,1\}, \quad 1 \le i, j \le 6 \quad . \tag{2.2.25}$$

Now consider BPSP$_2$:

$$f_2 = \min y \tag{2.2.26}$$

subject to

$$u_1, u_2, u_3 \le y \quad , \tag{2.2.27}$$

$$
\begin{aligned}
C_{11} + (\delta_{11} + \delta_{14}) &\le u_1 \quad , \\
C_{12} + (\delta_{12} + \delta_{15}) &\le u_1 \quad , \\
C_{13} + (\delta_{13} + \delta_{16}) &\le u_1 \quad ,
\end{aligned}
\tag{2.2.27}
$$

$$
\begin{aligned}
C_{21} + (\delta_{21} + \delta_{24} + \delta_{31} + \delta_{34} + \delta_{41} + \delta_{44}) &\le u_2 \quad , \\
C_{22} + (\delta_{22} + \delta_{25} + \delta_{32} + \delta_{35} + \delta_{42} + \delta_{45}) &\le u_2 \quad , \\
C_{23} + (\delta_{23} + \delta_{26} + \delta_{33} + \delta_{36} + \delta_{43} + \delta_{46}) &\le u_2 \quad ,
\end{aligned}
\tag{2.2.28}
$$

$$
\begin{aligned}
C_{31} + (\delta_{51} + \delta_{54} + \delta_{61} + \delta_{64}) &\le u_3 \quad , \\
C_{32} + (\delta_{52} + \delta_{55} + \delta_{62} + \delta_{65}) &\le u_3 \quad , \\
C_{33} + (\delta_{53} + \delta_{56} + \delta_{63} + \delta_{66}) &\le u_3 \quad ,
\end{aligned}
\tag{2.2.29}
$$

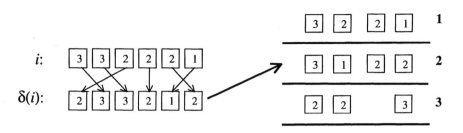

Figure 2.2.6. The input sequence and the content of the layers after the assignment

and (2.2.23)-(2.2.25).
BPSP₃:

$$f_3 = \min(u_1 + u_2 + u_3)$$

subject to (2.2.23)-(2.2.25) and (2.2.28)-(2.2.30).
For all these models, the permutation $\delta = (2, 1, 3, 6, 4, 5)$ is optimal. The objective functions f_1, f_2 and f_3 have the values 1, 2 and 4, respectively. Fig. 2.2.6 illustrates this example.

2.3. Complexity Results

THEOREM 2.6 *Problem D-BPSP₂ is NP-complete.*

Proof. It is easy to see that BPSP₂ ∈ NP, since a nondeterministic algorithm needs only to guess a permutation of the variables and to check in polynomial time whether that permutation satisfies all the given constraints. We proceed by showing that the *3-SAT (3-Satisfiability)* problem can be polynomially reduced to BPSP₂. Concerning the complexity issues of the 3-SAT problem we refer the reader to [32]. Below we give the definition of the Satisfiability problem [64].

DEFINITION 2.7 *Let $X = \{x_1, x_2, ..., x_n\}$ be a set of n Boolean variables. A literal y_i is either a variable x_i or its negation $\bar{x}_i$. A clause F_j is a disjunction of literals. Let formula $F = F_1 \wedge F_2 \wedge ... \wedge F_m$ be a conjunction of m clauses. The formula F is satisfiable if and only if there is a truth assignment $t : X \to \{0, 1\}$, which simultaneously satisfies all clauses F_j in F. The Satisfiability problem is the problem to decide for a given instance (X, F) whether there is a truth assignment for X that satisfies F. The 3-SAT problem is a restriction of the Satisfiability problem where each clause contains exactly 3 literals. The 3-SAT problem is still NP-complete.*

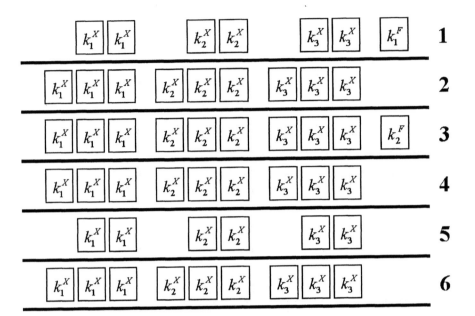

Figure 2.3.7. The content of the storage before distributing the objects from the sequence *seq*

For an arbitrary instance (X, F) of the *3-SAT* problem we define an instance of the sequencing problem $BPSP_2$, such that there exists a feasible permutation of the objects δ with

$$\max_{k=1,\ldots,N^K,\, l=1,\ldots,N^L} \left(C_{kl} + \sum_{i=1}^{N^O} \sum_{j=1}^{N^O} D_{ikjl} \delta_{ij} \right) \leq B$$

if and only if there is a truth assignment $t : X \rightarrow \{0, 1\}$ satisfying F.

We choose $B = 3$ and $N^S = 1$ (*i.e.*, an object can move forward at most by one position). The number of layers is

$$N^L := 2(m + 1) \quad , \tag{2.3.1}$$

the number of objects is

$$N^O := 2nN^L \quad , \tag{2.3.2}$$

and the number of sets $\mathcal{S}_k$ is

$$N^K := 2n + m + 2mn \quad . \tag{2.3.3}$$

Specifically, we have

- $2n$ colors k_i^X, k_i^L, one for each variable x_i,

- m colors k_j^F, one for each clause F_j,

- the first group of mn auxiliary colors k_{ij}^{XF}, and

- the second group of mn auxiliary colors $k_{ij}^{\bar{X}F}$.

The sequence seq consists of $2n$ subsequent parts seq_i of N^L objects each, i.e., $seq := (seq_1, seq_2, ..., seq_n, seq_{n+1}, seq_{n+2}, ..., seq_{2n})$. Each seq_i has $2(m+1)$ objects, i.e., $seq_i := (q_{i,1}, q_{i,2}, ..., q_{i,2m+2})$. Now we define the sets S_k, i.e., the color of each object contained in the sequence seq_i :

the first and last objects of seq_{2i-1} and seq_{2i} always have the colors of the variable x_i, i.e.,

$$q_{2i-1,1}, q_{2i,1} \in S_{k_i^X} \tag{2.3.4}$$

$$q_{2i-1,N^L}, q_{2i,N^L} \in S_{k_i^L}. \tag{2.3.5}$$

The colors of the $2m$ objects in between are defined depending on the occurrence of variable x_i in the clauses F_j as follows:

$$x_i \in F_j \implies \begin{array}{l} q_{2i,2j} \in S_{k_j^F} \\ q_{2i,2j+1} \in S_{k_{ij}^{\bar{X}F}} \\ q_{2i-1,2j}, q_{2i-1,2j+1} \in S_{k_{ij}^{XF}} \end{array} \tag{2.3.6}$$

$$\bar{x}_i \in F_j \implies \begin{array}{l} q_{2i-1,2j} \in S_{k_j^F} \\ q_{2i-1,2j+1} \in S_{k_{ij}^{XF}} \\ q_{2i,2j}, q_{2i,2j+1} \in S_{k_{ij}^{\bar{X}F}} \end{array} \tag{2.3.7}$$

$$x_i, \bar{x}_i \notin F_j \implies \begin{array}{l} q_{2i-1,2j}, q_{2i-1,2j+1} \in S_{k_{ij}^{XF}} \\ q_{2i,2j}, q_{2i,2j+1} \in S_{k_{ij}^{\bar{X}F}} \end{array} \tag{2.3.8}$$

The values of the cost function C_{kl} (which reflects the content of the storage) are defined for any $\{k, l\}$ as:

$$C_{k_i^X, l} = \begin{cases} 2, & \text{if } l \in \{1, N^L - 1\} \\ 3, & \text{else} \end{cases}, \forall i \in \{1, ..., n\} \tag{2.3.9}$$

$$C_{k_j^F, l} = \begin{cases} 1, & \text{if } l = 2j - 1 \\ 0, & \text{else} \end{cases}, \forall j \in \{1, ..., m\} \tag{2.3.10}$$

$$C_{k_{ij}^{XF}, l}, C_{k_{ij}^{\bar{X}F}, l}, C_{k_i^L, l} = 0, \forall i, j, l. \tag{2.3.11}$$

Fig. 2.3.7 illustrates these coefficients for an example with $n = 3$ and $m = 2$. In the remaining part of the proof we show that a feasible permutation of the objects in *seq* with

$$\max_{k=1,\ldots,N^K,\ l=1,\ldots,N^L} \left(C_{kl} + \sum_{i=1}^{N^O} \sum_{j=1}^{N^O} D_{ikjl} \delta_{ij} \right) \leq 3$$

exists if and only if F is satisfiable, *i.e.*, there is a truth assignment $t : X \rightarrow \{0,1\}$ that satisfies each clause in F.

First, we assume that F is satisfiable. We show that there exists a feasible permutation of objects with no more than 3 objects of the same color on any layer. Recall that:

- for each color k_i^X there are already exactly three objects on each of the layers $2, \ldots, N^L-2, N^L$, two objects on each of the layers $1, N^L-1$, one object in seq_{2i-1} and one object in seq_{2i};

- for each color k_j^F there is one object on layer $2j - 1$ and there are exactly three objects in *seq*;

- for each color k_{ij}^{XF}, $k_{ij}^{\bar{X}F}$ there exist at most two objects in *seq*;

- for each color k_i^L only two objects exist in seq_i.

Now consider the first object of subsequence seq_{2i-1}. This object has color k_i^X. It has to be sent to layer $N^L - 1$ or has to remain on the first layer, because all other layers already contain three objects of this color. The layers 1 and $N^L - 1$ already contain two objects of color k_i^X and can accommodate only one additional object each. Therefore, the second object of color k_i^X – the first object of seq_{2i} – has to be sent to the layer not used by the first object. This will be done by moving the objects to the nearest layer (*i.e.*, move of minimal distance), such that for any $j \in seq_i$ the following holds: $\delta(j) \in seq_i$. That means we can discuss each subsequence independently. We will use this alternation to map the truth assignments into the permutations set δ:

$$t(x_i) = 1 \Longrightarrow \begin{cases} \delta(q_{2i-1,1}) = q_{2i-1,N^L-1} & \\ \delta(q_{2i-1,j}) = q_{2i-1,j-1} & , \quad 1 < j < N^L \\ \delta(q_{2i-1,N^L}) = q_{2i-1,N^L} & \\ \delta(q_{2i,j}) = q_{2i,j} & , \quad 1 \leq j \leq N^L \end{cases} \quad (2.3.12)$$

$$t(x_i) = 0 \Longrightarrow \begin{cases} \delta(q_{2i,1}) = q_{2i,N^L-1} & \\ \delta(q_{2i,j}) = q_{2i,j-1} & , \quad 1 < j < N^L \\ \delta(q_{2i,N^L}) = q_{2i,N^L} & \\ \delta(q_{2i-1,j}) = q_{2i-1,j} & , \quad 1 \leq j \leq N^L \end{cases} \quad (2.3.13)$$

In this way, if t assigns 1 to x_i, then from (2.3.12) the first object of seq_{2i-1} moves to layer $N^L - 1$, and all other objects in seq_{2i-1} move one layer up (except for the last object which remains on the last layer). All objects of seq_{2i} keep their positions. Otherwise, if t assigns 0 to x_i, then it follows from (2.3.13) that all objects from seq_{2i-1} keep their positions and the first object of seq_{2i} moves to the layer $N^L - 1$. All other objects in seq_{2i} move one layer up (except for the last object which remains on the last layer).

Assume there is a layer l with four or more objects of the same color. Let us first discuss the color. It cannot be one of $k_{ij}^{XF}, k_{ij}^{\bar{X}F}, k_i^L$ since there are at most two of those and C_{kl} is zero for them. Without loss of generality we assume that there are four objects of color k_j^F on layer l. Since $|S_{k_l^F}| = 3$, $C_{k_j^F, l}$ must be 1 and because of (2.3.10) it follows that $l = 2j - 1$. Now consider one of the three objects. From (2.3.6) and (2.3.7) it is known that $S_{k_l^F} \subset \{q_{2i,2j}, q_{2i-1,2j}\}$. We distinguish two cases:

Case 1:

$$q_{2i,2j} \in S_{k_l^F} \overset{(2.3.6)}{\Longrightarrow} x_i \in F_j \wedge \delta(q_{2i,2j}) = q_{2i,2j-1} \qquad (2.3.13)$$

$$\text{since } l = 2j - 1 \overset{(2.3.13)}{\Longrightarrow} t(x_i) = 0 \qquad (2.3.14)$$

Case 2:

$$q_{2i-1,2j} \in S_{k_l^F} \overset{(2.3.6)}{\Longrightarrow} \bar{x}_i \in F_j \wedge \delta(q_{2i-1,2j}) = q_{2i-1,2j-1} \qquad (2.3.15)$$

$$\text{since } l = 2j - 1 \overset{(2.3.12)}{\Longrightarrow} t(\bar{x}_i) = 0 \qquad (2.3.16)$$

Therefore, in both cases the truth value of the literal of x_i in F_j is false, so F_j contains one false literal. The same reasoning holds for the other two objects. Thus, F_j contains three false literals and is therefore false itself. This is a contradiction to the assumption that t satisfies (X, F).

For the missing direction of the proof, we assume that (X, F) is unsatisfiable. Our goal is to prove that for any permutation δ there exists at least one color for which four objects are located on the same layer. Barring trivial cases, that will be color k_j^F.

Consider the permutations where $\delta(q_{2i,1}) \equiv l \bmod N^L$ with $2 \leq l < N^L - 1$ or $l = N^L$, i.e., when the first object of subsequence $2i$ is not moved to the first or last-but-one layer. Because of (2.3.9) $C_{k_i^X, l} = 3$, layer l contains four objects of color k_i^X. Therefore we only need to consider the remaining cases, i.e., when the first object moved to the first or to the next to last layer.

With the same reasoning we can assume that $\delta(q_{2i-1,1}) \equiv \pm 1 \bmod N^L$. So we only deal with cases where

$$\delta(q_{2i,1}) \equiv 1 \bmod N^L \text{ and } \delta(q_{2i-1,1}) \equiv -1 \bmod N^L \qquad (2.3.18)$$

or

$$\delta(q_{2i,1}) \equiv -1 \bmod N^L \text{ and } \delta(q_{2i-1,1}) \equiv 1 \bmod N^L \quad . \qquad (2.3.19)$$

Let us define a truth assignment t_δ by

$$t(x_i) = \begin{cases} 1 & \text{if } \delta(q_{2i,1}) \equiv 1 \bmod N^L \\ 0 & \text{if } \delta(q_{2i,1}) \equiv -1 \bmod N^L \end{cases} \quad . \qquad (2.3.20)$$

Furthermore, (2.2.1) tells us that $\delta(j) \geq j-1$ as $N^S = 1$ by construction. Since (X, F) is unsatisfiable by assumption, there must be some clause F_j which is not satisfied by t_δ.

Case 1: F_j contains a non-negated literal x_i.

Then $t_\delta(x_i) = 0$ and from (2.3.20) we know, that $\delta(q_{2i,1}) \equiv -1 \bmod N^L$ and from (2.3.19) that $\delta(q_{2i-1,1}) \equiv 1 \bmod N^L$. Because $\delta(q_{2i,1}) \geq q_{2i,1} - 1 \equiv 0 \bmod N^L$, so $\delta(q_{2i,1}) \geq q_{2i,1} + N^L - 2$ since $\delta(k) > k - 2$, i.e., the tray $q_{2i,1}$ stays in the additional storage at least until $N^L - 2$ objects passed. Therefore, $\delta(q_{2i,k}) = q_{2i,k} - 1$ for $2 \leq k \leq N^L - 1$, especially for $k = 2j$. From (2.3.6) we know that $q_{2i,2j} \in S_{k_j^F}$, hence there is one additional object of color k_j^F in layer $2j - 1$.

Case 2: F_j contains a negated literal $\bar{x}_i$.

Then $t_\delta(x_i) = 1$ and from (2.3.20) we know that $\delta(q_{2i,1}) \equiv 1 \bmod N^L$ and from (2.3.18) that $\delta(q_{2i-1,1}) \equiv -1 \bmod N^L$. As in the first case, we conclude that $\delta(q_{2i-1,1}) \geq q_{2i-1,1} + N^L - 2$ and, therefore, it follows that $\delta(q_{2i-1,2j}) = q_{2i-1,2j} - 1 \equiv 2j - 1 \bmod N^L$. Since (2.3.7) $q_{2i-1,2j} \in S_{k_j^F}$ there is an additional object of color k_j^F on layer $2j - 1$.

Hence, for each of the three literals of F_j there is one object of color k_j^F on layer $2j - 1$ and $C_{2j-1,k_j^F} = 1$ according to (2.3.10). Therefore, layer $2j - 1$ contains four objects of color k_j^F and the proof is complete.

To see that this transformation can be performed in polynomial time, it suffices to observe that the number of layers, the number of colors and the number of objects in *seq* are bounded by a polynomial in $2(m + 1)$, $4n(m + 1)$ and $2n + m + 2mn$, respectively. Hence the size of the BPSP$_2$ instance is bounded above by a polynomial function of the size of the 3-SAT instance (q.e.d). ∎

We illustrate Theorem 2.6 by the following example.

EXAMPLE 2.8 *Suppose there is the following instance of the 3-SAT problem: $F = (x_1 \vee x_2 \vee x_3) \wedge (\bar{x}_1 \vee x_2 \vee \bar{x}_3)$, i.e., $n = 3$ and $m = 2$. Now define* *all data needed for constructing the sequence seq. By (2.3.1)-(2.3.3) we obtain $N^L = 6$, $N^O = 36$, and the number of sets S_k is $N^K = 20$, i.e.,*

- k_i^X, k_i^L, $i = 1, 2, 3$;

- k_j^F, $j = 1, 2$;

- k_{ij}^{XF}, $k_{ij}^{\bar{X}F}$, $i = 1, 2, 3$, $j = 1, 2$.

Fig. 2.3.7 shows the content of the storage system w.r.t. formulas (2.3.9) and (2.3.10) before the objects from seq are distributed. Fig. 2.3.8 shows the sequence seq = $(seq_1, seq_2, ..., seq_6)$. In addition, this picture illustrates the assignments, e.g., for x_1, described by formulas (2.3.12) and (2.3.13) and the subsequent assignment to the layers. Similar transformations apply to the sequences seq_3, seq_4, seq_5, seq_6 (the first two correspond to x_2, the others to x_3).

2.4.   Polynomial Subcases

The decision problem D-BPSP$_2$ is shown to be NP-complete (see Section 2.3). What makes this problem difficult? In this section we consider the problem with some additional assumptions. The first assumption is that the numbers of elements in sets S_k, i.e., values $|S_k|$, are known in advance. The second one is that the capacity of the pre-sorting facility is large enough, i.e., given $N^S \geq N^L - 1$, any permutation of objects can be realized. This assumption is only needed for an alternative model formulation provided in Subsection 2.4.2.

2.4.1   Reformulation of BPSP$_2$ and BPSP$_3$

In Section 2.2.3.2 we suggested model formulations with continuous variables u_k. We can decrease the number of variables if the values of $|S_k|$, the number of objects of color k in the input sequence, are known in advance. Clearly, the optimal permutation of the objects is one that allows a uniform distribution of objects over all layers. Of course, we need to add to the value $|S_k|$ all objects of color k that are already in the layers $\sum_{l=1}^{N^L} C_{kl}$. It can happen that $\max\limits_{l=1,...,N^L} C_{kl} > \left\lceil \frac{|S_k| + \sum_{l=1}^{N^L} C_{kl}}{N^L} \right\rceil$. Therefore, we have to calculate values $S_k^* = \max\left(\left\lceil \frac{|S_k| + \sum_{l=1}^{N^L} C_{kl}}{N^L} \right\rceil, \max\limits_{l=1,...,N^L} C_{kl} \right)$.

Thus, the corresponding objective function value for BPSP$_3$ is $\sum_{k=1}^{N^K} u_k = \sum_{k=1}^{N^K} S_k^*$ and the equality $\max\limits_{k=1,...,N^K} u_k = \max\limits_{k=1,...,N^K} S_k^*$ holds for BPSP$_2$. Then we can modify the optimization models from Section 2.2.3.2 and obtain the transformed model:

D-BPSP$^/$:

$$\sum_{j=1}^{N^O} \delta_{ij} = 1, \quad 1 \le i \le N^O \quad , \tag{2.4.1}$$

$$\sum_{i=1}^{N^O} \delta_{ij} = 1, \quad 1 \le j \le N^O \quad , \tag{2.4.2}$$

$$C_{kl} + \sum_{i=1}^{N^O} \sum_{j=1}^{N^O} D_{ikjl}\delta_{ij} \le S_k^*, \quad \begin{matrix} 1 \le k \le N^K \\ 1 \le l \le N^L \end{matrix} \quad , \tag{2.4.3}$$

$$\delta_{ij} = 0, \quad \forall i, j : j < i - N^S \quad , \tag{2.4.4}$$

$$\delta_{ij} \in \{0, 1\}, \quad 1 \le i, j \le N^O \quad . \tag{2.4.5}$$

This is not yet an optimization problem since only a feasible point of (2.4.1)-(2.4.5) need to be found. If the capacity of the pre-sorting facility is not large enough, the feasible set can be empty.

EXAMPLE 2.9 *Here we demonstrate the calculation of* S_k^* *values on the data from Example 2.5:*

$$S_1^* = \max\left(\left\lceil \frac{|S_1| + C_{11} + C_{12} + C_{13}}{N^L} \right\rceil, \max_{j=1,2,3} C_{1j}\right) = \max\left(\left\lceil \frac{1+1+0+0}{3} \right\rceil, 1\right) = 1;$$

$$S_2^* = \max\left(\left\lceil \frac{|S_2| + C_{21} + C_{22} + C_{23}}{N^L} \right\rceil, \max_{j=1,2,3} C_{2j}\right) = \max\left(\left\lceil \frac{3+1+2+0}{3} \right\rceil, 2\right) = 2;$$

$$S_3^* = \max\left(\left\lceil \frac{|S_3| + C_{31} + C_{32} + C_{33}}{N^L} \right\rceil, \max_{j=1,2,3} C_{3j}\right) = \max\left(\left\lceil \frac{2+0+0+1}{3} \right\rceil, 1\right) = 1.$$

2.4.2   An Alternative Model Formulation of D-BPSP$^/$

In the following we show that D-BPSP$^/$ without (2.4.4) is polynomially solvable. At first, we present another mathematical model formulation named $D - \overline{BPSP^/}$ with the following integer variables:

x_{kl} is the number of objects of color k on layer l .

Notice that these variables can be derived easily from the δ_{ij} variables:

$$x_{kl} = \sum_{i=1}^{N^O} \sum_{j=1}^{N^O} D_{ikjl}\delta_{ij} \quad . \tag{2.4.6}$$

Suppose $N^O = pN^L$, where $p \in \mathbb{Z}$. Then the decision problem D-$\overline{\text{BPSP}'}$ concerns whether an assignment exists for x_{kl} satisfying the constraints:

$$\sum_{l=1}^{N^L} x_{kl} = |\mathcal{S}_k|, \quad 1 \leq k \leq N^K \quad , \tag{2.4.7}$$

$$\sum_{k=1}^{N^K} x_{kl} = \frac{N^O}{N^L}, \quad 1 \leq l \leq N^L \quad , \tag{2.4.8}$$

$$0 \leq x_{kl} \leq \mathcal{S}_k^* - C_{kl}, \quad 1 \leq k \leq N^K, \quad 1 \leq l \leq N^L \quad . \tag{2.4.9}$$

After that we show the correspondence between D-BPSP$'$ and D-$\overline{\text{BPSP}'}$ by demonstration that a solution of the first problem implies a solution of the second one and vice versa.

THEOREM 2.10 *Problems D-BPSP$'$ (without 2.4.4) and D-$\overline{\text{BPSP}'}$ are equivalent.*

Proof. *We show that if there is a solution of D-BPSP$'$ $\delta_{ij} \in \{0,1\}$, $1 \leq i,j \leq N^O$ that satisfies (2.4.1)-(2.4.3), then the solution x_{kl} of D-$\overline{\text{BPSP}'}$ satisfies (2.4.7)-(2.4.9), and vice versa.*

1:
$$\sum_{l=1}^{N^L} x_{kl} \overset{(2.4.6)}{=} \sum_{l=1}^{N^L} \sum_{i=1}^{N^O} \sum_{j=1}^{N^O} D_{ikjl}\delta_{ij} \overset{(2.2.10)}{=} \sum_{l=1}^{N^L} \sum_{i=1}^{N^O} \sum_{j=1}^{N^O} S_{ik}M_{jl}\delta_{ij}$$

$$= \sum_{i=1}^{N^O} S_{ik} \sum_{j=1}^{N^O} \delta_{ij} \sum_{l=1}^{N^L} M_{jl} \overset{(2.2.9)}{=} \sum_{i=1}^{N^O} S_{ik} \underbrace{\sum_{j=1}^{N^O} \delta_{ij}}_{=1 \text{ by } (2.4.1)}$$

$$= \sum_{i=1}^{N^O} S_{ik} \overset{(2.2.8)}{=} |\mathcal{S}_k|, \quad 1 \leq k \leq N^K \Longleftrightarrow \sum_{j=1}^{N^O} \delta_{ij} = 1, \quad 1 \leq i \leq N^O;$$

2:
$$\sum_{k=1}^{N^K} x_{kl} \overset{(2.4.6)}{=} \sum_{k=1}^{N^K} \sum_{i=1}^{N^O} \sum_{j=1}^{N^O} D_{ikjl}\delta_{ij} \overset{(2.2.10)}{=} \sum_{k=1}^{N^K} \sum_{i=1}^{N^O} \sum_{j=1}^{N^O} S_{ik}M_{jl}\delta_{ij}$$

$$= \sum_{j=1}^{N^O} M_{jl} \sum_{i=1}^{N^O} \delta_{ij} \sum_{k=1}^{N^K} S_{ik} \overset{(2.2.8)}{=} \sum_{j=1}^{N^O} M_{jl} \underbrace{\sum_{i=1}^{N^O} \delta_{ij}}_{=1 \text{ by } (2.4.2)}$$

$$= \sum_{j=1}^{N^O} M_{jl} \overset{(2.2.9)}{=} \frac{N^O}{N^L}, \quad 1 \leq l \leq N^L \Longleftrightarrow \sum_{i=1}^{N^O} \delta_{ij} = 1, \quad 1 \leq j \leq N^O;$$

3: *The constraints (2.4.3) and (2.4.9) are the same. Indeed,*

$$x_{kl} \leq \mathcal{S}_k^* - C_{kl} \Longleftrightarrow x_{kl} + C_{kl} \leq \mathcal{S}_k^* \overset{(2.4.6)}{\Longleftrightarrow} C_{kl} + \sum_{i=1}^{N^O} \sum_{j=1}^{N^O} D_{ikjl}\delta_{ij} \leq \mathcal{S}_k^*,$$

$$\forall k = \{1, ..., N^K\}, \quad \forall l = \{1, ..., N^L\} \ (q.e.d). \ \blacksquare$$

EXAMPLE 2.11 *Here we illustrate how to convert a solution of* $\overline{D\text{-}BPSP^{\prime}}$
to a solution of $D\text{-}BPSP^{\prime}$. *Consider the input sequence and sets*

$$\mathcal{S}_k = \{i_{c_k} \, | c_k = 1, \ldots, |\mathcal{S}_k|\}$$

from Example 2.5. Suppose a solution of $\overline{D\text{-}BPSP^{\prime}}$ *to be:*

$$
\begin{array}{lll}
x_{11} = 0, & x_{12} = 1, & x_{13} = 0, \\
x_{21} = 1, & x_{22} = 0, & x_{23} = 2, \\
x_{31} = 0, & x_{32} = 1, & x_{33} = 1.
\end{array}
$$

We construct the solution δ_{ij} *of* $D\text{-}BPSP^{\prime}$ *as follows:*
    For all $k = 1, ..., N^K$, $l = 1, ..., N^L$ *set initially* $c_k := 1$ *and* $i_l := 0$.
Then repeat the following steps while $x_{kl} \geq 1$:

- *let* $i^* := i_{c_k}$, $c_k := c_k + 1$;

- *let* $j^* := l + i_l N^L$, $i_l := i_l + 1$;

- *let* $\delta_{i^* j^*} := 1$ *and* $x_{kl} := x_{kl} - 1$.

Using this procedure for $N^K = 3$ *and* $N^L = 3$ *we obtain the solution:* $\delta = (2, 1, 3, 6, 5, 4)$. *Notice that this permutation yields the same assignment to the layers as in Example 2.5. But it is not feasible there, because the capacity of the pre-sorting facility is not large enough.*

Let us now show that the problem $\overline{D\text{-}BPSP^{\prime}}$ is polynomially solvable. We use the following results from [61].

THEOREM 2.12 *([61], Part III.1 "Totally unimodular matrices") Let* A *be a* $(0, 1, -1)$ *matrix with no more than two nonzero elements in each column. Then* A *is totally unimodular iff the rows of* A *can be partitioned into two subsets* A^1 *and* A^2 *such that if a column contains two nonzero elements, the following statements are true:*

1. *If both nonzero elements have the same sign, then one is in a row contained in* A^1 *and the other is in a row contained in* A^2.

2. *If the two nonzero elements have opposite sign, then both are in rows contained in the same subset.*

THEOREM 2.13 *([61], Part III.1 "Totally unimodular matrices") If* A *is totally unimodular, if* b, b', d *and* d' *are integral, and if* $P(b, b', d, d') =$

$\{x \in R^n : b' \le Ax \le b, \quad d' \le x \le d\}$ *is not empty, then* $P(b, b', d, d')$ *is an integral polyhedron.*

THEOREM 2.14 *The problem* $\overline{D\text{-}BPSP'}$ *is polynomially solvable.*
Proof. *We re-write (2.4.7)-(2.4.9) as:*

$$Ax = S^1_{N^K + N^L} \quad , \tag{2.4.10}$$

$$0 \le x \le S^2_{N^K N^L} \quad , \tag{2.4.11}$$

where

$$A = \begin{pmatrix} A^1 \\ A^2 \end{pmatrix} \quad , \tag{2.4.12}$$

$$A^1 = \begin{pmatrix} 1 & 1 & \dots & 1 & 0 & 0 & & 0 & & 0 & 0 & & 0 \\ 0 & 0 & & 0 & 1 & 1 & \dots & 1 & & 0 & 0 & & 0 \\ 0 & 0 & & 0 & 0 & 0 & & 0 & \dots & 0 & 0 & & 0 \\ 0 & 0 & & 0 & 0 & 0 & & 0 & & 1 & 1 & \dots & 1 \end{pmatrix} \quad ,$$

$$A^2 = \begin{pmatrix} 1 & 0 & & 0 & 1 & 0 & & 0 & \dots & 1 & 0 & & 0 \\ 0 & 1 & & 0 & 0 & 1 & & 0 & \dots & 0 & 1 & & 0 \\ 0 & 0 & \dots & 0 & 0 & 0 & \dots & 0 & \dots & 0 & 0 & \dots & 0 \\ 0 & 0 & & 1 & 0 & 0 & & 1 & \dots & 0 & 0 & & 1 \end{pmatrix} \quad ,$$

and

$$S^1_{N^K + N^L} = (|\mathcal{S}_1|, |\mathcal{S}_2|, ..., |\mathcal{S}_k|, \frac{N^O}{N^L}, \frac{N^O}{N^L}, ..., \frac{N^O}{N^L})^T,$$

$$S^2_{N^K N^L} = (S_1^* - C_{11}, ..., S_1^* - C_{1,N^L}, ..., S_{N^K}^* - C_{N^K, 1}, ..., S_{N^K}^* - C_{N^K, N^L})^T,$$

and

$$x = (x_{11}, x_{12}, ..., x_{1,N^L}, x_{21}, x_{22}, ..., x_{2,N^L}, ..., x_{N^K,1}, x_{N^K,2}, ..., x_{N^K,N^L})^T.$$

At first, notice that the matrix A *is totally unimodular:*
    By Theorem 2.12, the matrix A *is partitioned into two subsets (2.4.12), and each subset contains exactly one nonzero element of the same sign in each column.*
    Secondly, the polyhedron of (2.4.10)-(2.4.11) is integral by Theorem 2.13: $b' = b = S^1_{N^K + N^L}$, $d' = 0$, $d = S^2_{N^K N^L}$ *are integral, and obviously, the system of linear inequalities (2.4.10)-(2.4.11) always has a solution. And again from [61] it is well known that such problems are polynomially solvable (q.e.d).* ■

2.5.   The Case of Two Layers

In this section we analyze the BPSP with $N^L = 2$. We show that in this case the BPSP is polynomially solvable, and construct corresponding polynomial algorithms.

2.5.1 Offline Situations

In offline situations we know the arrival sequence of the N^O objects in advance. Consequently, the number $|S_k|$ of objects of type k is also known. It is obvious that the best solution is to assign half of objects of each type to layer 1 and the other half to layer 2. If $|S_k|$ is even, then there are exactly $\frac{|S_k|}{2}$ objects of the type k on each layer, otherwise there can be no more than $\left\lceil \frac{|S_k|}{2} \right\rceil$ objects. Thus, the optimal objective function value is

$$f(u) = \sum_{k=1}^{N^K} u_k := \sum_{k=1}^{N^K} \left\lceil \frac{|S_k|}{2} \right\rceil \quad , \qquad (2.5.1)$$

where u_k represents the maximal number of objects of type k over all layers (we use the objective function for BPSP$_3$). Without loss of generality we assume that N^O is even. The assignment of the objects can be realized as follows. We group each pair of two consecutive objects so that the whole sequence is divided into $\frac{N^O}{2}$ pairs. Let $\mathcal{L} = \{\mathcal{P}_1, \mathcal{P}_2, ..., \mathcal{P}_{\frac{N^O}{2}}\}$ be a list of these pairs. Obviously, in each pair only two types are presented, k_1 and k_2. Assume that in each pair $\mathcal{P}_i$ in $\mathcal{L}$ the first object with number $(2i-1)$ has type k_1 and the second object with number $2i$ belongs to type k_2. For each type k we define two functions:

- $p[k]$, that provides us with information about the objects assigned so far. This function takes values depending on the relation between the numbers of objects of type k on the first and second layer (M_{L_1} and M_{L_2}, respectively), *i.e.*,

$$p[k] = \begin{cases} -1, & M_{L_2}(k) = M_{L_1}(k) + 1 \\ 0, & M_{L_1}(k) = M_{L_2}(k) \\ 1, & M_{L_1}(k) = M_{L_2}(k) + 1 \end{cases} ;$$

- $d[k]$, that gives the number of objects of type k which are already assigned to the layers.

    Algorithm D_1 generates an assignment with the optimal objective function value for (2.5.1). The decision variables are defined in the following way:

$$\delta_{ij} = \begin{cases} 1, & \text{if object number } i \text{ is assigned to position number } j \\ 0, & \text{otherwise} \end{cases} .$$

The algorithm starts with the first pair of $\mathcal{L}$ and $\delta_{ij} := 0$ for all i and j. The objects from this first pair have types k_1 and k_2, respectively.

Let the object of type k_1 go to the first layer, and the object of type k_2 to the second. We save this assignment in $p[k_1]$ and $p[k_2]$ and delete this pair from $\mathcal{L}$. Then we choose the next pair according to the rule described in Step 4. The next time an object of type k_1 appears in a pair, we assign it to layer 2, and the other object of this pair to layer 1. We proceed as long as there are elements left over in $\mathcal{L}$.

Algorithm D_1

Steps of the algorithm:

1 Generate the list of pairs of objects $\mathcal{L} = \{\mathcal{P}_1, \mathcal{P}_2, ..., \mathcal{P}_{\frac{NO}{2}}\}$.

2 Calculate $p[k]$ for every k;

If all $p[k] = 0$ then goto Step 3, else goto Step 4.

3 Take the first element $\mathcal{P}_i$ from $\mathcal{L}$;

Determine the types k_1 and k_2 of objects $2i - 1$, $2i$, respectively;

Goto Step 6.

4 Take the first element $\mathcal{P}_i$ from $\mathcal{L}$ for which the following condition holds: $p[k_1] \neq 0$ or $p[k_2] \neq 0$ or both, where k_1 and k_2 are the types of the objects $2i - 1$ and $2i$, respectively.

5 If $p[k_1] = \begin{cases} +1, & \text{then goto Step 6,} \\ -1, & \text{then goto Step 7,} \\ 0, & \text{then goto Step 8.} \end{cases}$

6 Assign $\delta_{2i-1,2i} := \delta_{2i,2i-1} := 1$, *i.e.*, change the order of objects;

Goto Step 9.

7 Assign: $\delta_{2i-1,2i-1} := \delta_{2i,2i} := 1$, *i.e.*, keep the order of objects;

Goto Step 9.

8 If $p[k_2] = \begin{cases} +1, & \text{then goto Step 7,} \\ -1, & \text{then goto Step 6.} \end{cases}$

9 Change $p[k_1]$ and $p[k_2]$ correspondingly to the assignment on Step 6 or Step 7;

Delete $\mathcal{P}_i$ from $\mathcal{L}$, goto Step 2.

Note that the above algorithm does not consider the cases $p[k_1] = p[k_2] = 1$ and $p[k_1] = p[k_2] = -1$. As the following lemma shows, these cases do not occur.

LEMMA 2.15 *After completing each step of algorithm D_1 the following holds: for any n such that $\mathcal{P}_n \in \mathcal{L}$ and for any type k there exist at most two $p[k] \neq 0$. If $\exists k_1, k_2 : p[k_1] \neq 0$, $p[k_2] \neq 0$, then either $p[k_1] = 1$ and $p[k_2] = -1$, or $p[k_1] = -1$ and $p[k_2] = 1$, and, additionally, for all $k \neq k_1$, $k \neq k_2 : p[k] = 0$.*
   Proof:

Proof. We prove by induction over $n = 1, ..., \frac{N^O}{2}$.
   $n = 1$. This means that we start with the objects with number 1 and 2 and $p[k] = 0$, $k = 1, ..., N^K$.
   Consider the following cases:

1 Suppose the objects have different types. Let $1 \in \mathcal{S}_{k_1}$, $2 \in \mathcal{S}_{k_2}$ (the case $2 \in \mathcal{S}_{k_1}$, $1 \in \mathcal{S}_{k_2}$ is analogous). The first object goes to the first layer, *i.e.*, we assign $p[k_1] := 1$ in Step 8, the second one goes to the second layer, *i.e.*, we assign $p[k_2] := -1$ in Step 8. All other values $p[k]$ remain unchanged, *i.e.*, $p[k] = 0$, $k \neq k_1, k_2$.

2 Suppose both objects have the same type. Let $1, 2 \in \mathcal{S}_{k_1}$ (the case $1, 2 \in \mathcal{S}_{k_2}$ is analogous). After Step 8: $p[k_1] = 0 + 1 - 1 = 0$. All other values $p[k]$ are left unchanged, *i.e.*, $p[k] = 0$, $k \neq k_1$. So, we have $p[k] = 0$ for any k.

Thus, for $n = 1$ the statement is true. Now let us consider the induction step. Suppose the assumption is true for $n - 1$. Let us evaluate $p[k]$ for the case n:

1 The objects have different types k_1 and k_2. Depending on step $n - 1$ there are the following six possibilities:

   - $p[k_1] = 1$; $p[k] = 0$, $k \neq k_1$.
     After Step 8: $p[k_1] = 1 - 1 = 0$; $p[k_2] = 0 + 1 = 1$; $p[k] = 0$, $k \neq k_2$;
   - $p[k_1] = -1$; $p[k] = 0$, $k \neq k_1$.
     After Step 8: $p[k_1] = -1 + 1 = 0$; $p[k_2] = 0 - 1 = -1$; $p[k] = 0$, $k \neq k_2$;
   - $p[k_1] = 1$, $p[k_2] = -1$; $p[k] = 0$, $k \neq k_1, k_2$.
     After Step 8: $p[k_1] = 1 - 1 = 0$; $p[k_2] = -1 + 1 = 0$; So, $\forall k : p[k] = 0$;
   - $p[k_1] = -1$, $p[k_2] = 1$; $p[k] = 0$, $k \neq k_1, k_2$.
     After Step 8: $p[k_1] = -1 + 1 = 0$; $p[k_2] = 1 - 1 = 0$; $p[k] = 0$; So, $\forall k : p[k] = 0$;
   - $p[k_1] = 1$, $\exists k_3 : p[k_3] = -1$; $p[k_2] = 0$; $p[k] = 0$, $k \neq k_1$, $k \neq k_3$.

After Step 8: $p[k_1] = 1 - 1 = 0$; $p[k_2] = 0 + 1 = 1$; $p[k_3] = -1 + 0 = -1$; $p[k] = 0$, $k \neq k_2$, $k \neq k_3$;

- $p[k_1] = -1$, $\exists k_3 : p[k_3] = 1$; $p[k_2] = 0$; $p[k] = 0$, $k \neq k_1$, $k \neq k_3$.
  After Step 8: $p[k_1] = -1 + 1 = 0$; $p[k_2] = 0 - 1 = -1$; $p[k_3] = 1 + 0 = 1$; $p[k] = 0$, $k \neq k_2$, $k \neq k_3$.

These variants are symmetric with respect to k_1 and k_2 in the sense that we can replace k_1 by k_2 and vice versa. So, one can see that in any case we have no more then two values k with $p[k] \neq 0$. If we have exactly two such k, $p[k_1]$ and $p[k_2]$ have different signs.

2 Both objects have the same type k_1. Depending on the previous information during step n there are four possibilities:

- $p[k_1] = 1$; $p[k] = 0$, $k \neq k_1$.
  After Step 8: $p[k_1] = 1 - 1 + 1 = 1$; $p[k] = 0$, $k \neq k_1$;
- $p[k_1] = -1$; $p[k] = 0$, $k \neq k_1$.
  After Step 8: $p[k_1] = -1 + 1 - 1 = -1$; $p[k] = 0$, $k \neq k_1$;
- $p[k_1] = 1$, $p[k_2] = -1$; $p[k] = 0$, $k \neq k_1$, $k \neq k_2$.
  After Step 8: $p[k_1] = 1 - 1 + 1 = 1$; $p[k_2] = -1 + 0 = -1$; $p[k] = 0$, $k \neq k_1$, $k \neq k_2$;
- $p[k_1] = -1$, $p[k_2] = 1$; $p[k] = 0$, $k \neq k_1$, $k \neq k_2$.
  After Step 8: $p[k_1] = -1 + 1 - 1 = -1$; $p[k_2] = 1 + 0 = 1$; $p[k] = 0$, $k \neq k_1$, $k \neq k_2$.

Again, these variants are symmetric with respect to k_1 and k_2. Thus, in any case we have no more then two k: $p[k] \neq 0$. In case we have exactly two k, $p[k_1]$ and $p[k_2]$ have different signs (q.e.d).

∎

The following theorem ensures that the algorithm D_1 finds an optimal solution in polynomial time.

THEOREM 2.16 *Algorithm D_1 constructs an optimal solution in polynomial time.*
Proof. *The optimality of the generated assignment follows directly from algorithm D_1 and the lemma above, because we always assign the objects in such a way that one half of the objects of type k is assigned to section 1 and the second half to section 2. This is possible as long as we have the pre-sorting facility of capacity one. Let us see how many operations are necessary to construct the optimal solution. All steps of algorithm D_1 are repeated at most $\frac{N}{2}$ times, because each time we delete one element*

from $\mathcal{L}$. *Initially,* $\mathcal{L}$ *contains* $\frac{N}{2}$ *elements. In order to check if* $i \in S_k$ *we need no more than* N^K *operations. Checking if all* $p[k] = 0$ *requires also* N^K *operations. And finally, Step 3 can be repeated at most* $\frac{N}{2} - 3$ *times, because* i *has a value at least two before the step. So, in total there are* $O(N^O, N^K) = \frac{N^O}{2}(N^K + N^K + \frac{N^O}{2} - 3) = \frac{(N^O)^2}{4} + N^O N^K - \frac{3N^O}{2}$. *As one can see it is polynomial in the number of objects* N^O *as well as in the number of types* N^K *(q.e.d.).* ■

2.5.2   Online Situations

In the following we present an online algorithm that is linear in N^O and N^K. It is, therefore, much faster than the offline algorithm D_1. The sequence of objects appears in pairs: $\mathcal{P}_1, \mathcal{P}_2, \ldots, \mathcal{P}_{N^O/2}$. When we assign layers for the objects from $\mathcal{P}_i$, the online algorithm does not have any knowledge of $\mathcal{P}_j$, $j > i$. Only after $\mathcal{P}_i$ has been served, the next pair $\mathcal{P}_{i+1}$ becomes known. In total there are N^O objects. Without loss of generality, N^O is assumed to be even. We group two consecutively objects into $\frac{N^O}{2}$ pairs. Obviously, in each pair of objects there are no more than two types k_1 and k_2. Let C_{kl} specify the number of objects of type k already assigned to the layer l.

Algorithm D_2

For $i := 1$ to $\frac{N^O}{2}$ do:

1 Determine types k_1 and k_2 of the objects $2i - 1$, $2i$ respectively.

2 If $(k_1 = k_2)$ or $(C_{k_1 1} = C_{k_1 2})$ and $(C_{k_2 1} = C_{k_2 2})$, then goto Step 4, else goto Step 3.

3 If $(C_{k_1 1} > C_{k_1 2})$ or $(C_{k_1 1} = C_{k_1 2}$ and $C_{k_2 2} > C_{k_2 1})$, then goto Step 5, else goto Step 4.

4 Assign $\delta_{2i-1,2i-1} := \delta_{2i,2i} := 1$, *i.e.*, keep the order of objects unchanged;

   Goto Step 6.

5 Assign $\delta_{2i-1,2i} := \delta_{2i,2i-1} := 1$, *i.e.*, change the order of objects;

   Goto Step 6.

6 Change $C_{k_1 1}$, $C_{k_1 2}$, $C_{k_2 1}$ and $C_{k_2 2}$ with respect to the assignment.

   End For.

THEOREM 2.17 *The algorithm* D_2 *finds the best online solution in polynomial time.*

Proof. *We prove the theorem by induction over* $n = 1, ..., \frac{N}{2}$.
$n = 1$, *i.e*, $C_k^0 = 0$, $k = 1, ..., N^K$.

$$f^1(u) = \sum_{k=1}^{N^K} u_k^1 = \sum_{k=1}^{N^K} \max\{C_{k1}^0 + \sum_{i \in S_k} \delta_{i1}; C_{k2}^0 + \sum_{i \in S_k} \delta_{i2}\}$$

Consider the following cases:
1. *Both objects have the same type:* $\{1, 2\} \subseteq S_{k_1}$.

$$f^1(u) = \sum_{k=1}^{N^K} u_k^1 =$$

$$= \sum_{k=1}^{N^K} \max\{C_{k1}^0 + \sum_{i \in S_k} \delta_{i1}; C_{k2}^0 + \sum_{i \in S_k} \delta_{i2}\}$$

$$= \max\{C_{k_11}^0 + \sum_{i \in S_{k_1}} \delta_{i1}; C_{k_12}^0 + \sum_{i \in S_{k_1}} \delta_{i2}\}$$

$$+ \max\{C_{k_21}^0 + \sum_{i \in S_{k_2}} \delta_{i1}; C_{k_22}^0 + \sum_{i \in S_{k_2}} \delta_{i2}\}$$

$$= \max\{C_{k_11}^0 + \delta_{11} + \delta_{21}; C_{k_12}^0 + \delta_{12} + \delta_{22}\} + \max\{C_{k_21}^0; C_{k_22}^0\}$$

$$= \max\{\delta_{11} + \delta_{21}; \delta_{12} + \delta_{22}\} = 1,$$

Therefore, the objective function value is 1, independent of how we permute the objects. Only one and exactly one element of the sums $\delta_{11} + \delta_{21}$ *and* $\delta_{12} + \delta_{22}$ *has to be equal to 1. Consequently,* $\delta_{11} + \delta_{12} = \delta_{21} + \delta_{22} = 1$.

2. *The case* $\{1, 2\} \subseteq S_{k_2}$ *is analogous to Case 1.*
3. *The objects have different types:* $\{1\} \subseteq S_{k_1}$, $\{2\} \subseteq S_{k_2}$ $(k_1 \neq k_2)$.

$$f^1(u) = \sum_{k=1}^{N^K} u_k^1$$

$$= \sum_{k=1}^{N^K} \max\{C_{k1}^0 + \sum_{i \in S_k} \delta_{i1}; C_{k2}^0 + \sum_{i \in S_k} \delta_{i2}\}$$

$$= \max\{C_{k_11}^0 + \sum_{i \in S_{k_1}} \delta_{i1}; C_{k_12}^0 + \sum_{i \in S_{k_1}} \delta_{i2}\}$$

$$+ \max\{C_{k_21}^0 + \sum_{i \in S_{k_2}} \delta_{i1}; C_{k_22}^0 + \sum_{i \in S_{k_2}} \delta_{i2}\}$$

$$= \max\{C_{k_11}^0 + \delta_{11}; C_{k_12}^0 + \delta_{12}\} + \max\{C_{k_21}^0 + \delta_{21}; C_{k_22}^0 + \delta_{22}\}$$

$$= \max\{\delta_{11}; \delta_{12}\} + \max\{\delta_{21}; \delta_{22}\} = 1 + 1 = 2.$$

Therefore, the objective function value is 2, independent of how we permute the objects. Only one and exactly one element of the pairs $\{\delta_{11}; \delta_{12}\}$ *and* $\{\delta_{21}; \delta_{22}\}$ *has to be equal 1. Consequently,*

$$\max\{\delta_{11}; \delta_{12}\} = \max\{\delta_{21}; \delta_{22}\} = 1.$$

4. *The case* $\{2\} \subseteq S_{k_1}$, $\{1\} \subseteq S_{k_2}$ *is symmetric to Case 3.*

Thus, for $n = 1$ we get the minimal objective function value. Now consider the induction step. Suppose, the assumption is true for $n - 1$. Let us now analyze the case n. Again, we have to consider several cases:

1. Both objects are of the same type: $\{2n - 1, 2n\} \subseteq S_{k_1}$.

$$f^n(u) = \sum_{k=1}^{N^K} u_k^n$$

$$= \sum_{k=1}^{N^K} \max\{C_{k1}^{n-1} + \sum_{i \in S_k} \delta_{i1}; C_{k2}^{n-1} + \sum_{i \in S_k} \delta_{i2}\}$$

$$= \sum_{k=1, k \neq k_1, k_2}^{N^K} \max\{C_{k1}^{n-1} + \sum_{i \in S_k} \delta_{i1}; C_{k2}^{n-1} + \sum_{i \in S_k} \delta_{i2}\}$$

$$+ \max\{C_{k_1 1}^{n-1} + \sum_{i \in S_{k_1}} \delta_{i,2n-1}; C_{k_1 2}^{n-1} + \sum_{i \in S_{k_1}} \delta_{i,2n}\}$$

$$+ \max\{C_{k_2 1}^{n-1} + \sum_{i \in S_{k_2}} \delta_{i,2n-1}; C_{k_2 2}^{n-1} + \sum_{i \in S_{k_2}} \delta_{i,2n}\}$$

$$= \max\{C_{k_1 1}^{n-1} + \delta_{2n-1,2n-1} + \delta_{2n,2n-1}; C_{k_1 2}^{n-1} + \delta_{2n-1,2n} + \delta_{2n,2n}\}$$

$$+ \max\{C_{k_2 1}^{n-1}; C_{k_2 2}^{n-1}\} + \sum_{k=1, k \neq k_1, k_2}^{N^K} u_k^{n-1}$$

$$= u_{k_1}^{n-1} + 1 + u_{k_2}^{n-1} + \sum_{k=1, k \neq k_1, k_2}^{N^K} u_k^{n-1}$$

$$= f^{n-1}(u) + 1.$$

The objective function value increases by 1, independent of the permutation of the objects. Cf. Case 1 in the initial step of the induction.

2. The case $\{2n - 1, 2n\} \subseteq S_{k_2}$ is analogous to Case 1.

3. The objects have different types: $\{2n - 1\} \subseteq S_{k_1}$, $\{2n\} \subseteq S_{k_2}$.

$$f^n(u) = \sum_{k=1}^{N^K} u_k^n$$

$$= \sum_{k=1}^{N^K} \max\{C_{k1}^{n-1} + \sum_{i \in S_k} \delta_{i1}; C_{k2}^{n-1} + \sum_{i \in S_k} \delta_{i2}\}$$

$$= \sum_{k=1 | k \neq k_1, k_2}^{N^K} \max\{C_{k1}^{n-1} + \sum_{i \in S_k} \delta_{i1}; C_{k2}^{n-1} + \sum_{i \in S_k} \delta_{i2}\}$$

$$+ \max\{C_{k_1 1}^{n-1} + \sum_{i \in S_{k_1}} \delta_{i,2n-1}; C_{k_1 2}^{n-1} + \sum_{i \in S_{k_1}} \delta_{i,2n}\}$$

$$+ \max\{C_{k_2 1}^{n-1} + \sum_{i \in S_{k_2}} \delta_{i,2n-1}; C_{k_2 2}^{n-1} + \sum_{i \in S_{k_2}} \delta_{i,2n}\}$$

$$= \max\{C_{k_1 1}^{n-1} + \delta_{2n-1,2n-1}; C_{k_1 2}^{n-1} + \delta_{2n-1,2n}\}$$

$$+ \max\{C_{k_2 1}^{n-1} + \delta_{2n,2n-1}; C_{k_2 2}^{n-1} + \delta_{2n,2n}\} + \sum_{k=1 | k \neq k_1, k_2}^{N^K} u_k^{n-1}$$

In this case the objective function value depends on the permutation of the objects. Consider all possible combinations and calculate the objective function value after changing the order of the objects, or if the order of objects was kept:

		Change	*Keep*
$C_{k_1 1}^{n-1} > C_{k_1 2}^{n-1}$	$C_{k_2 1}^{n-1} > C_{k_2 2}^{n-1}$	$f^{n-1}(u)+1$	$f^{n-1}(u)+1$
	$C_{k_2 1}^{n-1} = C_{k_2 2}^{n-1}$	$f^{n-1}(u)+1$	$f^{n-1}(u)+2$
	$C_{k_2 1}^{n-1} < C_{k_2 2}^{n-1}$	$f^{n-1}(u)$	$f^{n-1}(u)+2$
$C_{k_1 1}^{n-1} = C_{k_1 2}^{n-1}$	$C_{k_2 1}^{n-1} > C_{k_2 2}^{n-1}$	$f^{n-1}(u)+2$	$f^{n-1}(u)+1$
	$C_{k_2 1}^{n-1} = C_{k_2 2}^{n-1}$	$f^{n-1}(u)+2$	$f^{n-1}(u)+2$
	$C_{k_2 1}^{n-1} < C_{k_2 2}^{n-1}$	$f^{n-1}(u)+1$	$f^{n-1}(u)+2$
$C_{k_1 1}^{n-1} < C_{k_1 2}^{n-1}$	$C_{k_2 1}^{n-1} > C_{k_2 2}^{n-1}$	$f^{n-1}(u)+2$	$f^{n-1}(u)$
	$C_{k_2 1}^{n-1} = C_{k_2 2}^{n-1}$	$f^{n-1}(u)+2$	$f^{n-1}(u)+1$
	$C_{k_2 1}^{n-1} < C_{k_2 2}^{n-1}$	$f^{n-1}(u)+1$	$f^{n-1}(u)+1$

Keeping the order implies $\delta_{2n-1,2n-1} := \delta_{2n,2n} := 1$,
Changing the order implies $\delta_{2n-1,2n} := \delta_{2n,2n-1} := 1$;

Note that algorithm D_2 assigns the values to the decision variables in such a way that the objective function is minimized, i.e., the value for $f^n(u)$ is minimal as long as $f^{n-1}(u)$ is minimal (assumption of the induction step).

4. The case $\{2n\} \subseteq S_{k_1}$, $\{2n-1\} \subseteq S_{k_2}$ is symmetric to Case 3. Thus, instead of keeping we need to change the order of the objects. This is what algorithm D_2 does.

Let us now see how many operations are necessary to construct the optimal solution. For checking if $i \in S_k$ we need no more than N^K operations, since in the worst case the set S_k can contain all input objects. Steps 1 to 6 are repeated $\frac{N^O}{2}$ times. Thus, in total we have $O(N^O) = \frac{N^O}{2} N^K$. As one can see, it is polynomial with respect to the number of objects N^O as well as to the number of types N^K (q.e.d). ■

2.5.3 Algorithms for Online Situations with Lookahead

2.5.3.1 Definition of a Lookahead

Very often the knowledge of future requests would help to improve the objective function value and to construct more efficient algorithms. Such situations will be referred as situations with lookahead. In recent years, a lot of problems with lookahead have been studied (paging problems, bin packing problems, graph problems and so on). It was expected that knowing a lookahead will improve the solution for the BPSP as well. The reason is the following: suppose we have the input sequence described in Example 2.23. Obviously, in this case, if we knew the third pair while serving the second - we will obtain a better solution. So, in this section we describe algorithms with lookahead for the BPSP with $N^L = 2$ (later we make a brief comparative analysis of algorithms with lookahead and without).

Formal definitions of lookahead can be found in [3], [38]. In this book we use the simplified definitions given in [29]:

DEFINITION 2.18 *Weak lookahead with respect to the number of requests (WLNR) The online algorithm sees the present request at time t and the next $\lambda - 1$ succeeding requests. Request $t + \lambda$ is not seen by the algorithm at time t. However, once the request $\delta(t)$ is processed, a new request, i.e., $\delta(t + \lambda)$, becomes known.*

DEFINITION 2.19 *Strong lookahead with respect to the number of requests (SLNR) The online algorithm sees λ present requests at time t, and before all these are processed no new request becomes known.*

In our case, we consider a lookahead containing not simple requests (objects), but pairs of objects, *i.e.,* we say that the size of the lookahead is $\lambda = n$ if the algorithm sees the present pair at time t and sees and considers n succeeding pairs.

2.5.3.2   An Algorithm with a Weak Lookahead, size $\lambda = 1$

The idea of this algorithm is quite simple: the sequence of objects is represented in pairs as was described above (for algorithm D_2). At the time we have to serve the current pair $\mathcal{P}_i$, also objects from pair $\mathcal{P}_{i+1}$ are known. It means there are four objects (two in each pair) and there are four possibilities to permute the objects: to keep the order of objects in both pairs, to change the order in both pairs and to keep the order of objects in one pair, changing the order in the second one. Then, depending on the values of the evaluated objective function $z = \sum_{k=1}^{N^K} \sum_{l=1}^{N^L} C_{kl}$, we realize the most efficient (z is minimal) permutation on objects from the current pair $\mathcal{P}_i$. Next we consider the following pair of objects, $\mathcal{P}_{i+1}$, and produce the same calculations. For the last pair, $\mathcal{P}_{\frac{N^O}{2}}$, we take the decision based on the previous assignment.

Algorithm WL$_1$

Initialize $i := 1$.

1 Determine the types k_1 and k_2 of the objects $2i - 1$, $2i$, respectively.

2 If $i < \frac{N^O}{2}$, then goto Step 3, else goto Step 5.

3 Determine the types l_1 and l_2 of the objects $2i+1$, $2i+2$, respectively;

Goto Step 4.

4 If $k_1 \neq l_1, l_2$ and $k_2 \neq l_1, l_2$, then goto Step 5, else goto Step 6.

5 For the current i, k_1 and k_2 apply Steps 2-6 of algorithm D_2;
   If $i = \frac{NO}{2}$, then STOP, otherwise Goto 15.

6 Assign temporarily $\delta_{2i-1,2i-1} := \delta_{2i,2i} := 1$, $\delta_{2i+1,2i+1} := \delta_{2i+2,2i+2} :=$
   1, *i.e.*, keep the order of both: the first pair of objects and the second
   pair of objects.

7 Calculate the value of $S_K^K = \sum_k \sum_l C_{kl}$.

8 Assign temporarily $\delta_{2i-1,2i-1} := \delta_{2i,2i} := 1$, $\delta_{2i+1,2i+2} := \delta_{2i+2,2i+1} :=$
   1, *i.e.*, keep the order of objects for the first pair and change the order
   of objects for the second pair.

9 Calculate the value of $S_C^K = \sum_k \sum_l C_{kl}$.

10 Assign temporarily $\delta_{2i-1,2i} := \delta_{2i,2i-1} := 1$, $\delta_{2i+1,2i+1} := \delta_{2i+2,2i+2} :=$
   1, *i.e.*, keep the order of objects for the second pair and change the
   order of objects for the first pair.

11 Calculate the value of $S_K^C = \sum_k \sum_l C_{kl}$.

12 Assign temporarily $\delta_{2i-1,2i} := \delta_{2i,2i-1} := 1$, $\delta_{2i+1,2i+2} := \delta_{2i+2,2i+1} :=$
   1, *i.e.*, to change the order of both: the first pair of objects and the
   second pair of objects.

13 Calculate the value of $S_C^C = \sum_k \sum_l C_{kl}$.

14 Compare the values of $S_K^K, S_C^K, S_K^C, S_C^C$. If S_K^K or S_C^K has the minimal
   value, then finally assign $\delta_{2i-1,2i-1} := \delta_{2i,2i} := 1$, otherwise assign
   $\delta_{2i-1,2i} := \delta_{2i,2i-1} := 1$.

15 $i := i + 1$;
   Goto Step 1.

2.5.3.3 Algorithms with a Weak Lookahead, size $\lambda \geq 2$

If we have a lookahead of size $\lambda \geq 2$ there are two different strategies to
handle the rest of the input sequence with number of pairs in lookahead
less or equal than λ. The first possibility: when we consider pair $\mathcal{P}_\lambda$,
such that $\lambda = \frac{NO}{2} - i$ to run the offline algorithm D_1 for the rest of
the sequence (algorithm WL_λ^1); the second possibility: to serve each

individual pair separately irrespective of how many pairs are left in the lookahead (algorithm WL_λ^2).

    Algorithm WL_λ^1

1 For $i := 1$ to $\frac{NO}{2}$ do:

2 If $\lambda = \frac{NO}{2} - i$ then apply the offline algorithm D_1 to the input sequence $\mathcal{P}_i, ..., \mathcal{P}_{i+\lambda}$ and *STOP*, else goto the next step.

3 Call algorithm D_1 with $\mathcal{P}_{i+1}, ..., \mathcal{P}_{i+\lambda}$ as input sequence.

4 Assign temporarily $\delta_{2i-1,2i-1} := \delta_{2i,2i} := 1$, *i.e.*, keep the order of objects in $\mathcal{P}_i$ unchanged.

5 Calculate the value of $S_K = \sum_k \sum_l C_{kl}$.

6 Assign temporarily $\delta_{2i-1,2i} := \delta_{2i,2i-1} := 1$, *i.e.*, change the order of objects in $\mathcal{P}_i$.

7 Calculate the value of $S_C = \sum_k \sum_l C_{kl}$.

8 If $S_K < S_C$ then assign constantly $\delta_{2i-1,2i-1} := \delta_{2i,2i} := 1$, else assign constantly $\delta_{2i-1,2i} := \delta_{2i,2i-1} := 1$.

    End For.

    Algorithm WL_λ^2

    For $i := 1$ to $\frac{NO}{2}$ do:

1 If $\lambda < (\frac{NO}{2} - i)$, then goto Step 2, else goto Step 3.

2 Apply algorithm D_1 to the input sequence $\mathcal{P}_{i+1}, ..., \mathcal{P}_{i+\lambda}$ and goto Step 6.

3 If $\frac{NO}{2} - i = 0$, then goto Step 4, else goto Step 5.

4 Apply algorithm D_2 to the last pair $\mathcal{P}_i$ and *STOP*.

5 Apply algorithm D_1 to the input sequence $\mathcal{P}_{i+1}, ..., \mathcal{P}_{\frac{NO}{2}}$ and goto Step 6.

6 Assign temporarily $\delta_{2i-1,2i-1} := \delta_{2i,2i} := 1$, *i.e.*, keep the order of objects in $\mathcal{P}_i$ unchanged.

7 Calculate the value of $S_K = \sum_k \sum_l C_{kl}$.

8 Assign temporary $\delta_{2i-1,2i} := \delta_{2i,2i-1} := 1$, *i.e.*, change the order of objects in $\mathcal{P}_i$.

9 Calculate the value of $S_C = \sum_k \sum_l C_{kl}$.

10 If $S_K < S_C$ then finally assign $\delta_{2i-1,2i-1} := \delta_{2i,2i} := 1$, else assign $\delta_{2i-1,2i} := \delta_{2i,2i-1} := 1$.

End For.

2.5.3.4 An Algorithm with a Strong Lookahead

In the following we present an online algorithm with strong lookahead of size λ. It means that at the beginning pairs $\mathcal{P}_1, ..., \mathcal{P}_\lambda$ are known, after serving them pairs $\mathcal{P}_{\lambda+1}, ..., \mathcal{P}_{2\lambda}$ become known, and so on. Therefore, we divide the input sequence into $q := \lceil \frac{N O}{\lambda} \rceil$ groups and operate with each group individually.

Algorithm SL$_\lambda$

For $i := 1$ to q do:
Apply the offline algorithm D_1 to the input sequence $\mathcal{P}_{(i-1)\lambda+1}, ..., \mathcal{P}_{i\lambda}$.
End For.

2.5.4 Competitive Analysis

The efficiency of online algorithms can be evaluated using competitive analysis, where online algorithms are compared with an optimal offline algorithm. First we collect some definitions and terms used throughout this chapter. One basic concept used for estimating the performance of online algorithms is the *competitive ratio*. The competitive ratio of an online algorithm can be defined with respect to an optimal offline algorithm, which knows an input sequence in advance and produces an optimal solution with minimal cost. Let A be an online algorithm, δ is an input sequence of requests and $f_A(\delta)$ is the cost achieved by A on the input sequence δ. Denote by $f_{OPT}(\delta)$ the cost achieved by an optimal offline algorithm OPT on the same input. Now we can give the definition of the competitive ratio of the online algorithm A.

DEFINITION 2.20 *([11])An online algorithm A is $c-$competitive if there is a constant α such that for all finite input sequences δ,*

$$f_A(\delta) \le c \cdot f_{OPT}(\delta) + \alpha.$$

When for the additive constant $\alpha \le 0$ holds, we may say for emphasis that A is strictly $c-$competitive.

In the following the competitiveness of the algorithms from Section 2.5 is analyzed. We will show that algorithm D_2 is strictly $\frac{3}{2}-$ competitive. To do this we have to prove the following two lemmas.

LEMMA 2.21 *Any online algorithm which solves the given sequencing problem is strictly N^L- competitive, where N^L is the number of layers.*
Proof. *The objective function has the largest value if all objects of the same type are assigned to the same layer. Thus, the maximum number, u_k, of objects of type k in any layer is $|S_k|$. Evaluate the objective function value: $f(u) = \sum\limits_{k=1}^{N^K} u_k = \sum\limits_{k=1}^{N^K} |S_k|$. Clearly, the corresponding optimal offline objective function value is $f_{D_1}(u) = \sum\limits_{k=1}^{N^K} u_k = \sum\limits_{k=1}^{N^K} \left\lceil \frac{|S_k|}{N^L} \right\rceil$. Suppose, that for all $k : |S_k| \equiv 0 \mod N^L$, then $\left\lceil \frac{|S_k|}{N^L} \right\rceil = \frac{|S_k|}{N^L}$. If there exists k such that $|S_k| \equiv \{1, \ldots, N^L - 1\} \mod N^L$, then $\left\lceil \frac{|S_k|}{N^L} \right\rceil > \frac{|S_k|}{N^L}$ and*

$$\frac{f(u)}{f_{D_1}(u)} = \frac{\sum\limits_{k=1}^{N^K} |S_k|}{\sum\limits_{k=1}^{N^K} \left\lceil \frac{|S_k|}{N^L} \right\rceil} \leq \frac{\sum\limits_{k=1}^{N^K} |S_k|}{\sum\limits_{k=1}^{N^K} \frac{|S_k|}{N^L}} \leq \frac{\sum\limits_{k=1}^{N^K} |S_k|}{\frac{1}{N^L} \sum\limits_{k=1}^{N^K} |S_k|} \leq N^L \quad .$$

This implies that the competitive ratio is N^L (q.e.d). ■

Consider a situation with only two layers, *i.e.*, $N^L = 2$. Then the following lemma holds.

LEMMA 2.22 *If an algorithm A is a $c-$competitive online algorithm, then $\frac{3}{2} \leq c \leq 2$.*
Proof. *From Lemma 2.21 with $N^L = 2$ we conclude that the upper bound for c is 2. We show that c cannot be less than $\frac{3}{2}$. For any positive integer number G we define $K := 4G$, $N := 8G$ and $|S_k| = 2$ for all $k = 1, \ldots, N^K$, i.e., we only have two objects of each type. Consider the following input sequence $S = S_1 S_2 \ldots S_G$, where S_g $(g = 1, \ldots, G)$ consists of eight elements. $S_g = \{8g - 7, \ldots, 8g - 4, 8g - 3, \ldots, 8g\}$. Let $8g - 7 \in S_{4g-3}$; $8g - 6 \in S_{4g-2}$; $8g - 5 \in S_{4g-1}$; $8g - 4 \in S_{4g}$. Define the types for the next four objects $\{8g - 3, \ldots, 8g\}$ depending on how algorithm A operates. Any online algorithm (with the same number of known objects) gets the input sequence of objects in pairs. The algorithm has to decide whether to keep the order of objects or to change it. If the algorithm keeps the order, the first object goes to the first layer, the second one to the second. Otherwise, the first object is sent to the second*

layer, and the second one to the first layer. Hence, after serving the first two pairs $\{8g-7, 8g-6\}$ *and* $\{8g-5, 8g-4\}$ *we can get four cases (see Table 2.1) and depending on the decision we can construct the next two pairs of objects:* $\{8g-3, 8g-2\}$ *and* $\{8g-1, 8g\}$ *for* S_g *(see Table 2.2):*

Table 2.1: *Results sequences after running the algorithm A*

1	$8g-7 \in S_{4g-3}; 8g-6 \in S_{4g-2};$	$8g-5 \in S_{4g-1}; 8g-4 \in S_{4g};$
2	$8g-7 \in S_{4g-3}; 8g-6 \in S_{4g-2};$	$8g-5 \in S_{4g}; 8g-4 \in S_{4g-1};$
3	$8g-7 \in S_{4g-2}; 8g-6 \in S_{4g-3};$	$8g-5 \in S_{4g-1}; 8g-4 \in S_{4g};$
4	$8g-7 \in S_{4g-2}; 8g-6 \in S_{4g-3};$	$8g-5 \in S_{4g}; 8g-4 \in S_{4g-1};$

Then an algorithm has to serve the next 2 pairs: $\{8g-3, 8g-2\}, \{8g-1, 8g\}$.

Table 2.2: *Recommended sequences for the next objects:* $\{8g-3, \ldots, 8g\}$

1	$8g-3 \in S_{4g-3}; 8g-2 \in S_{4g-1};$	$8g-1 \in S_{4g-2}; 8g \in S_{4g};$
2	$8g-3 \in S_{4g-3}; 8g-2 \in S_{4g};$	$8g-1 \in S_{4g-2}; 8g \in S_{4g-1};$
3	$8g-3 \in S_{4g-2}; 8g-2 \in S_{4g-1};$	$8g-1 \in S_{4g-3}; 8g \in S_{4g};$
4	$8g-3 \in S_{4g-2}; 8g-2 \in S_{4g};$	$8g-1 \in S_{4g-3}; 8g \in S_{4g-1};$

Let us determine the objective function value after serving these pairs. Regardless of the algorithm's decision, the maximum number, u_k, *of objects in any layer has the following values. Consider case 1:*

If an algorithm changes the order of $\{4g-3, 4g-1\}$, *then* $u_{4g-3} = 1$, $u_{4g-1} = 2$; *otherwise* $u_{4g-3} = 2$, $u_{4g-1} = 1$. *If an algorithm changes the order of* $\{4g-2, 4g\}$, *then* $u_{4g-2} = 1$, $u_{4g} = 2$; *otherwise* $u_{4g-3} = 2$, $u_{4g-1} = 1$. *Thus, for* S_g *we have* $f_{A_{S_g}}(u) = u_{4g-3} + u_{4g-1} + u_{4g-2} + u_{4g} = 3 + 3 = 6$. *We get the same result for the remaining cases. Since objects from different* S_g *have different types, we obtain*

$$f_A(u) = \sum_{g=1}^{G} f_{A_{S_g}}(u) = 6G = 6\frac{N^K}{4} = \frac{3}{2}N^K \quad .$$

At the same time the optimal objective function value is

$$f(u) = \sum_{k=1}^{N^K} u_k = \sum_{k=1}^{N^K} \frac{N_k}{2} = \sum_{k=1}^{N^K} 1 = N^K \quad .$$

This means $c = \frac{3}{2}$ *(q.e.d.)*. ∎

Let us illustrate Lemma 2.22 by the following example.

EXAMPLE 2.23

If $G = 1 \implies N^K = 4$, $N^O = 8$; $|\mathcal{S}_k| = 2$ for $k = 1, ..., 4$, we have to construct the input sequence $S = S_1$ where S_1 consists of eight objects. The first four objects have types $1, 2, 3, 4$, respectively. After executing the algorithm we get one of the following cases:

- the algorithm kept the order of objects for both pairs. Therefore, we get the output sequence: $1, 2, 3, 4$. In this case the next four objects (from 5 to 8) must have types $1, 3, 2, 4$, respectively;

- the algorithm changed the order of the first pair and kept the order of the second. We get: $2, 1, 3, 4$. In this case the next four objects (from 5 to 8) must have types $2, 3, 1, 4$;

- the algorithm kept the order of the first pair and changed the order of the second. We get: $1, 2, 4, 3$. In this case the next four objects (from 5 to 8) must have types $1, 4, 2, 3$;

- the algorithm changed the order of both pairs. We have: $3, 1, 4, 2$. In this case the next four objects (from 5 to 8) must have types $3, 4, 1, 3$.

For the first case, independently of the actions of the algorithm on the objects $5, 6, 7, 8$, after these actions the following statements hold:

Both objects of the same type, either 1 or 3 are on the same layer;

Both objects of the same type, either 2 or 4 are on the same layer.

An equivalent argumentation holds for all other cases. Therefore, the best value of the objective function for the *online* case is

$$f_A(u) = \sum_{k=1}^{4} u_k = 1 + 2 + 1 + 2 = 6 \quad . \tag{2.5.2}$$

At the same time we get the optimal value of the objective function for the *offline* case:

$$f(u) = \sum_{k=1}^{4} u_k = \sum_{k=1}^{4} \frac{|\mathcal{S}_k|}{2} = 4 \quad . \tag{2.5.3}$$

Using the Lemmas 4 and 5 we can now prove the following theorem.

THEOREM 2.24 *The algorithm D_2 is strictly $\frac{3}{2}$-competitive.*
Proof. *The algorithm D_2 operates as follows. At first, the sequence of all incoming objects is grouped into a list of pairs of objects. These pairs*

are numbered from 1 through $\frac{N^O}{2}$. From each pair of objects with types k_1 and k_2, the object of type k_1 is assigned to section L_1 and the object of type k_2 is assigned to section L_2. Thus, the next time an object of type k_1 appears in a pair, it is assigned to section L_2 and the other object of the current pair to section L_1. A problem could only occur if both objects of the current pair should be assigned to the same section, but this is not possible. This case was already considered in the proof of Lemma 2.22. Therefore the competitive ratio of algorithm D_2 is $\frac{3}{2}$ (q.e.d). ∎

2.5.5 Comparison of D_2 and Online Algorithms with Lookahead

Here we show that additional knowledge of input data improves the objective function value. Consider the worst case for the algorithm D_2, i.e., the following input sequence $(1, 2, 3, 4, 1, 3, 2, 4)$ (the construction rule was described in Lemma 2.22 and Example 2.23). In Table 2.3, the objective function values, $f(i)$, for offline and online algorithms are presented, as well as the resulting permutations $\delta(i)$.

Table 2.3: Comparison of offline and online algorithms

alg.	λ	$f(i)$	$\delta(i)$
D_1	0	4	$(1, 2, 4, 3, 6, 5, 7, 8)$
D_2	0	6	$(1, 2, 3, 4, 6, 5, 7, 8)$
WL_1	1	4	$(1, 2, 4, 3, 6, 5, 7, 8)$
WL_λ^1	1	6	$(2, 1, 3, 4, 5, 6, 7, 8)$
	2	4	$(1, 2, 4, 3, 6, 5, 7, 8)$
	3	4	$(1, 2, 4, 3, 6, 5, 7, 8)$
WL_λ^2	1	4	$(2, 1, 3, 4, 5, 6, 8, 7)$
	2	6	$(2, 1, 4, 3, 6, 5, 8, 7)$
	3	6	$(2, 1, 4, 3, 6, 5, 8, 7)$
SL_λ	1	6	$(1, 2, 3, 4, 5, 6, 7, 8)$
	2	6	$(1, 2, 3, 4, 5, 6, 7, 8)$
	3	4	$(1, 2, 4, 3, 6, 5, 7, 8)$

Thus, with only one pair of objects in the lookahead algorithms WL_1, WL_λ^2 produce an optimal permutation. Algorithms WL_λ^1 and SL_λ require the knowledge of two and three pairs, respectively, to get an optimal permutation.

2.6. Extensions

In this section we consider a specific case of the sequencing problem, in which the number of layers is a power of 2, i.e., 2^P. Accordingly, without loss of generality, the number of known objects, N^O, is divided by 2^P.

We show that for this problem there exists a polynomial algorithm if the capacity N^S of the pre-sorting facility is sufficiently large.

In this case we can use the algorithm D_2 consequently 2^{P-1} times. In the following, we refer to this modification as algorithm D_2^P.

     Algorithm D_2^P

Apply the algorithm D_2 to

- the whole sequence of N^O objects.

  Output: two subsequences of $\frac{N^O}{2}$ objects;

- each of two subsequences of $\frac{N^O}{2}$ objects.

  Output: four subsequences of $\frac{N^O}{4}$ objects;

     ... ...

- each of 2^{P-1} subsequences of $\frac{N^O}{2^{P-1}}$ objects.

  Output: 2^P subsequences of $\frac{N^O}{2^P}$ objects.

THEOREM 2.25 *The D_2^P algorithm described above is strictly $\left(\frac{3}{2}\right)^P$-competitive.*

Proof. *Follows directly from Theorem 2.24 and the description of D_2 (q.e.d).* ■

THEOREM 2.26 *The algorithm D_2^P constructs a solution in polynomial time.*

Proof. By Theorem 2.17 the complexity of D_2 is $O(N^O, N^K) = \frac{N^O}{2}N^K$. Now we calculate the complexity of D_2^P. The first step of D_2^P costs $\frac{N^O}{2}N^K$ operations, the second - $2\frac{N^O}{2^2}N^K$. Then we run D_2 four times for the $\frac{N^O}{2^2}$ objects, and thus the number of operations is $4\frac{N^O}{2^3}N^K$ and so on. In total the complexity is:

$$O(N^O, N^K) \;=\; \frac{N^O N^K}{2} + 2\frac{N^O N^K}{2^2} + ... + 2^{P-1}\frac{N^O N^K}{2^P}$$

$$=\; \sum_{p=1}^{P} 2^{p-1}\frac{N^O N^K}{2^p} = \frac{N^O N^K}{2}P \quad (q.e.d) \quad .$$

■

As soon as a polynomial algorithm for three layers will be constructed, it is possible to develop an algorithm similar to D_2^P for the problems with number of layers 3^P, and $2^{P_1}3^{P_2}$.

2.7.   Summary und Future Research

In this chapter we developed mathematical formulations and integer programming models for the Batch PreSorting Problems (BPSP). The main result is the proof of NP-completeness. So far it is still open whether $BPSP_3$ is polynomial or NP-complete. However, we expect that $BPSP_3$ is NP-complete as well. As the complexity proof for $BPSP_2$ involved the SAT problem, it is useful to adapt some existing efficient algorithms that have been developed to solve the SAT problem to the $BPSP_2$ problem.

In addition, we have investigated what makes this problem difficult. We constructed an alternative formulation and showed that it has an integer polyhedron and is equivalent to a restricted variant of $BPSP_2$ and $BPSP_3$; thus, this restricted version is also polynomially solvable. In addition we considered a polynomial subcase of the original formulation - the case of two layers. For this special case, we constructed one exact offline algorithm and several online algorithms with and without lookahead. We compared them using competitive analysis. One of those algorithm was adapted for a more general case of 2^P layers.

For further research, we suggest to construct a similiar polynomial algorithm for the case of three layers, and then to modify it for the problem with $2^{P_1} 3^{P_2}$ layers in online case. For a better estimation of the efficiency of such algorithms it would be helpful to get a non-trivial (different from a constant) lower bound for a general BPSP. Another possible research direction is to construct a transformation from the problem with two layers/many colors to the problem with many layers/two colors.

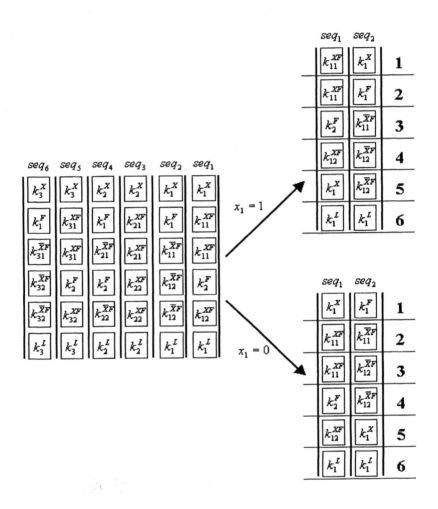

Figure 2.3.8. Changes in the sequence after assigning a value to x_1

Chapter 3

BATCH PRESORTING PROBLEMS. II APPLICATIONS IN INVENTORY LOGISTICS

In this chapter we consider a concrete application of BPSP to the storage system Rotastore, which is introduced in Section 3.1. Numerical results obtained when we applied our BPSP algorithms and models to the Rotastore system, are summarized in Section 3.2.

3.1. The Storage System Rotastore

In the following we describe the storage system Rotastore, developed and produced by psb GmbH, Pirmasens, Germany [73].

3.1.1 A Brief Description of the Rotastore

This system overcomes the poor performance of stacker crane systems by implementing a modular multi-carousel principle. This principle allows parallel loading and unloading by means of elevators interfacing the vertically staked carousel layers. As shown in Fig. 3.1.1, a typical configuration consists of two elevators and a stack of carousels each with identical numbers of slots. Identically sized objects are transported to and from the lifts by conveyors. Each carousel and each elevator may move independently. As indicated by the arrows, two combs of punches push objects simultaneously on all layers in or out of the carousel stack. The whole assembly is controlled by a sophisticated high-level programmable computer system which keeps track of operations and inventory.

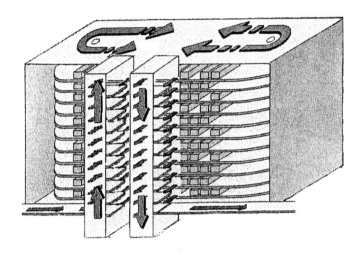

Figure 3.1.1. Typical Rotastore configuration with elevators

The system normally operates in an output-driven mode, *i.e.*, an input/output-cycle (I/O-cycle) starts with accepting a set of orders with one ordered object on each layer. The carousels then place, by rotation, the ordered objects in front of the output elevator ("seeking"). Next, the punches push out all ordered objects onto the output elevator which immediately starts unloading them onto the leaving conveyer ("stepping"). The carousels move the now empty slots in front of the input elevator and the punch push in the waiting objects. While the elevators load and unload, the next I/O-cycle starts with seeking the next set of orders. The expected performance of a Rotastore with m slots on each of n layers was investigated by Hamacher *et al.* [39]. In Subsection 3.1.2 most important results from their work are reviewed.

3.1.2 Stochastic Measures

For the performance of the system, the number of I/O-cycles per time unit is of course critical, therefore in [39] the expected duration of an I/O-cycle under uniform assumptions was computed. For the following analysis it is assumed that the outgoing conveyor is fast enough, never blocked and there are always objects waiting for input. These assumptions can be met by installing sufficient buffer space before and after the elevators.

THEOREM 3.1 *([39])Let the orders be independently uniformly distributed. Then the expected duration of an I/O cycle of a Rotastore with n layers is*

$$Et_{IO} = \frac{n + b^{n+1}}{n+1} \max(t_{mseek}, t'_{elev}) + 2t_{push} + t_{shift} \quad ,$$

where t_{mseek} is the maximal time needed to seek an object on one layer, t_{elev} is the time the elevators need for stepping, $t'_{elev} = t_{elev} - t_{push} - t_{shift}$ is the part of the stepping time which happens in parallel to seeking, t_{push} and t_{shift} are the times for pushing and shifting, and finally, $b := \min\left(\frac{t'_{elev}}{t_{mseek}}, 1\right)$ is the ratio between stepping and seeking time.

The same result has been derived for a combined input/output elevator (paternoster) by Kartnig [51] and Klinger [54]. The ratio b then simplifies to $b := \frac{t_{elev}}{t_{mseek}}$, since the elevator starts stepping only after the second push of the cycle.

The next result, derived by Hamacher *et al.* is related to the expected number of the occupied layers. It is assumed that the Rotastore accepts more than n orders which are fulfilled at a time. The assumption of one independent order, uniformly distributed over the layers is generalized to n_T independent uniformly distributed orders anywhere in the Rotastore, where n_T may be larger than n.

THEOREM 3.2 *([39])Let there be a set of n_T independent uniformly distributed orders. Then the expected number of occupied layers is*

$$En_{occ} = \sum_{k=0}^{n} \sum_{i=0}^{k} (-1)^i \binom{k}{n} \binom{k}{i} \binom{m(k-i)}{n_T} \binom{nm}{n_T}^{-1} (n-k) \quad .$$

Direct computation of En_{occ} shows, that for practically relevant cases one can expect between 50% and 80% of the layers to be occupied by orders. Practical experience shows that additional measures have to be taken into account to distribute the objects eventually with respect to the layers in order to realize the parallelizing speedup of the Rotastore. We will consider the case of a distribution center of a mayor German department store company. In this case, the objects are already partitioned into sets of orders when input into the Rotastore and the Rotastore may output a complete set of orders as soon as all objects of the set of orders are present. To improve the performance, the incoming objects are pre-sorted such that each batch spreads over as many layers as possible. Conveyors transport the objects throughout the system. Reordering is done online by stacker frames. A stacker frame is a small robotic device that lifts one object from the conveyor to let an arbitrary number of

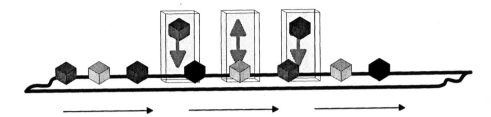

Figure 3.1.2.　Three stacker frames

objects pass. As Fig. 3.1.2 illustrates, usually several stacker frames are placed in a row along the conveyer so an object may overtake several preceding objects easily. The assembly is usually located above the input accumulating conveyor (*e.g.*, roller conveyor) directly in front of the Rotastore.

3.2.　Numerical Tests

Because we can formulate the problem of minimizing the number of I/O cycles of the Rotastore as the BPSP, we can apply all solution approaches developed in Chapter 2. First, we check which model is more efficient, *i.e.*, gives the best performance of the Rotastore. Second, we compare offline and online algorithms for the polynomial case when the Rotastore has only two layers, *i.e.*, $N^L = 2$.

3.2.1　Models

As was already mentioned, all proposed BPSP model formulations can be used to model the Rotastore. Thus, in this section we test the models $BPSP_1$ to $BPSP_3$. Computational experiments were carried out with randomly generated examples of the sets S_k using ILOG's OPL-studio [47]. Input data were chosen using real-world data knowledge: the usual number of objects in the input sequence is about 900, the Rotastore has eight layers of 113 slots each. We do not take in consideration the number of slots, because the layers are occupied only about 50% - 80% as mentioned above. The input of Rotastore could be rearranged by two stacker frames. Since the average size of the sets of objects is ten, the number of different orders is about 90. Thus, using the notations introduced in Section 2.2 we set:

$$
\begin{aligned}
N &= 900 && \text{(the total number of objects);} \\
N^O &= 60 && \text{(the number of known objects);} \\
N^K &= 90 && \text{(the number of colors);} \\
N^L &= 8 && \text{(the number of layers);} \\
N^S &= 2 && \text{(the number of stacker frames).}
\end{aligned}
$$

In order to distribute all objects among the layers of the Rotastore (each time only N^O objects are available, the next N^O objects become known only after distribution of the first N^O objects) it is necessary to solve the model $\lceil N/N^O \rceil$ times. The input sequence is generated as N^K sets S_k. For example, if $N^K = 4$, $N^O = 7$, then the generated sets could be:

$$
S_1 = \{1,4\}, \quad S_2 = \{\emptyset\}, \quad S_3 = \{2,5,6,7\}, \quad S_4 = \{3\} \quad .
$$

We want to observe the number, N^{OC}, of output cycles for the Rotastore using the solutions of the models BPSP$_1$ and BPSP$_2$. Obviously, BPSP$_3$ gives a minimal value for N^{OC}, because the objective function of BPSP$_3$ (2.2.21) minimizes this directly. As in BPSP$_3$, the variables u_k are related to the output cycles in BPSP$_2$, therefore $N^{OC} = \sum_{k=1}^{N^K} u_k$. In BPSP$_1$ we do not have those variables, but they can be calculated indirectly as follows: $u_k = \max_{l=1,\ldots,N^L} R_{kl}$, where R_{kl} is a number of objects of color k on layer l after assigning of all objects to the layers of the Rotastore. For instance, for the input sequence in Example 2.5, $R_{11} = 1$, $R_{12} = 1$, $R_{13} = 0$, $R_{21} = 2$, $R_{22} = 2$, $R_{23} = 2$, $R_{31} = 1$, $R_{32} = 1$, $R_{33} = 1$ (see Fig. 2.2.6). Thus $u_1 = 1$, $u_2 = 2$, $u_3 = 1$ and $N^{OC} = 4$.

We summarize the results of the experiments for BPSP$_1$ and BPSP$_2$ in tables such as Table 3.1 and Table 3.2, respectively (further result tables are collected in Appendix A.A.1). Values in column "opt N^{OC}" are calculated by BPSP$_3$ and Δ is the optimality gap, which measures how far a solution is away from the optimum for the current cycle :

$$
\Delta = \frac{N^{OC} - opt\,N^{OC}}{N^{OC}} \cdot 100\% \quad . \tag{3.2.1}
$$

Table 3.1: Results for BPSP$_1$

cycle	N^S	cpu, sec.	N^{OC}	opt N^{OC}	$\Delta, \%$
1	2	2.45	45	44	2.27
2	2	2.34	75	68	10.29
3	2	2.38	93	79	17.72
4	2	2.31	107	86	24.42
5	2	2.15	124	88	40.91
6	2	2.16	140	92	52.17
7	2	2.17	154	98	57.14
8	2	2.15	175	103	69.90
9	2	2.20	187	112	66.96
10	2	2.23	199	120	65.83
11	2	2.40	210	128	64.06
12	2	2.19	225	134	67.91
13	2	2.19	242	142	70.42
14	2	2.22	254	152	67.11
15	2	2.22	262	161	62.73

Table 3.2: Results for BPSP$_2$

cycle	N^S	cpu, sec.	N^{OC}	opt N^{OC}	$\Delta, \%$
1	2	10.82	44	44	0.00
2	2	7.03	68	68	0.00
3	2	6.75	79	79	0.00
4	2	6.81	85	86	-1.16
5	2	6.75	87	88	-1.14
6	2	7.20	99	92	7.61
7	2	7.14	109	98	11.22
8	2	6.82	117	103	13.59
9	2	6.81	129	112	15.18
10	2	6.70	135	120	12.50
11	2	6.71	142	128	10.94
12	2	6.98	148	134	10.45
13	2	6.86	161	142	13.38
14	2	6.81	170	152	11.84
15	2	15.60	181	161	12.42

For each value N^S, ten runs were made (each one starting with an empty Rotastore, *i.e.*, at the beginning all $C_{kl} = 0$) with different inputs. Figures 3.2.3 and 3.2.4 show the average values of CPU time (in seconds) and the number of output cycles, respectively.

It was observed that BPSP$_1$ and BPSP$_2$ produce solutions which are on average worse by 54% and 7% than those obtained by BPSP$_3$, respectively. The CPU time for BPSP$_1$ is usually about 2 seconds and for BPSP$_2$ is 7 seconds. Similar results were obtained for different values of $N^S = [1, ..., N^L]$. For BPSP$_3$ the CPU time varies between 8 seconds

$(N^S = [3, ..., N^L])$, 10-15 seconds ($N^S = 1$), and above 200,000 seconds $(N^S = 2)$.

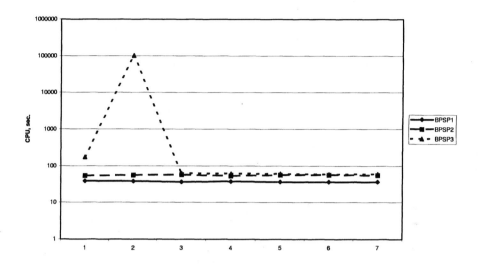

Figure 3.2.3. The average CPU time versus the number of stacker frames

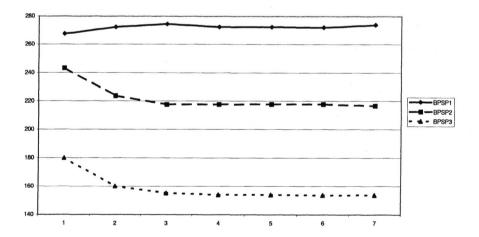

Figure 3.2.4. Average number of I/O cycles versus the number of stacker frames

The CPU time seems to depend on the degree of filling of the Rotastore (see, for instance, Fig. 3.2.5). On average, one observes that

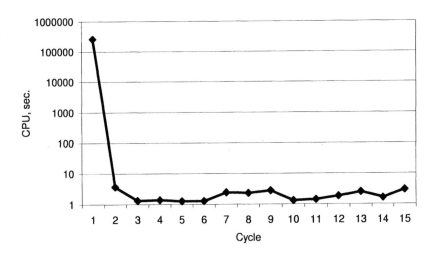

Figure 3.2.5. CPU time for BPSP$_3$ with $N^S = 2$

the CPU time decreases when the Rotastore contains more objects (C_{kl} becomes larger). This is explained by the fact that the feasible region becomes smaller, and therefore, there are less possible combinations of the binary variables δ_{ij}. As a consequence of this, the number of active nodes in the branch-and-bound algorithm becomes smaller, that in turn decreases the running time. So, we recommend to use BPSP$_3$ when the Rotastore is not empty. However, in practical situations this happens very seldom. In some rare cases, when the Rotastore is empty, we can fill it using BPSP$_2$, and continue with BPSP$_3$. In this case, the computed solution is near-optimal ($\Delta \approx 1 - 2\%$) and the CPU time is drastically decreased. Sometimes the combination of BPSP$_2$ and BPSP$_3$ (for the first cycle to run BPSP$_2$ and for the cycles 2-14 to run BPSP$_3$) can produce better solutions than applying BPSP$_3$ repeatedly 15 times. Consider, for instance, the data in Table 3.3. For cycle 6, both BPSP$_2$ and BPSP$_3$, produced solutions with the same objective function value 92, but the solutions differ *w.r.t.* the assignment of the objects to the layers. Thus, before cycle 7 starts, these models have different initial data and, therefore, the combination of the models produced, after 15 cycles, solutions with better results in shorter time (compare the CPU time displayed on Fig. 3.2.5 with CPU time displayed on Table 3.3) than applying BPSP$_3$ 15 times.

Table 3.3: Results using the combination of BPSP_2 and BPSP_3

cycle	N^S	cpu, sec.	N^{OC}	opt N^{OC}	$\Delta, \%$
1	2	10.82	44	44	0.00
2	2	21.07	68	68	0.00
3	2	2.51	79	79	0.00
4	2	2.52	86	86	0.00
5	2	2.54	88	88	0.00
6	2	2.61	92	92	0.00
7	2	2.59	97	98	-0.01
8	2	3.04	102	103	-0.01
9	2	2.82	107	112	-0.04
10	2	2.84	115	120	-0.04
11	2	2.70	124	128	-0.03
12	2	3.24	134	134	0.00
13	2	2.54	141	142	-0.01
14	2	3.49	150	152	-0.01
15	2	2.91	158	161	-0.02

In addition, we observe that it is not useful to have more than four stacker frames because this does not lead to a smaller number of output cycles. This fact was also confirmed for different input sequences.

3.2.2 Algorithms

In this section we present computational results for the algorithms described in Section 2.5.3.1. In order to compare the efficiency of those algorithms we implemented them using the programming language C++. At first, we want to estimate the quality of the solutions produced by the online algorithm D_2. Secondly, we check the influence of the lookahead on those solutions. The size of the lookahead, λ, *i.e.*, the number of pairs known in advance, varies between one and five pairs. Finally, we investigate how often the competitive ratio of $\frac{3}{2}$ is reached on randomly generated example. For $N^O = 900$, $N^K = 90$ the result of the comparative analysis of all algorithms compared to the exact offline algorithm D_1 is given in Table 3.4; N^{OC} is the number of output cycles needed for complete output; Δ is the optimality gap defined in Section 3.2.1. More results of the experiments are presented in Appendix A.A.2; note that the CPU time is not reported there, because it always was smaller than 6 seconds.

Table 3.4: Comparative analysis of all algorithms

alg.	λ	N^{OC}	$\Delta, \%$	cpu, sec.
D_1		472		0
D_2	0	481	1.91	0
WL_1	1	480	1.67	0
WL_λ^1	1	479	1.48	5
	2	479	1.48	3
	3	479	1.48	1
	4	486	2.97	2
	5	486	2.97	3
WL_λ^2	1	477	1.06	2
	2	478	1.27	4
	3	478	1.27	5
	4	484	2.54	3
	5	483	2.33	5
SL_λ	1	559	18.43	1
	2	555	17.58	1
	3	557	18.01	5
	4	554	17.37	1
	5	553	17.16	5

Table 3.4 shows that algorithm D_2 is very efficient and produces near-optimal objective function values, *i.e.*, the competitive ratio $c = 1.5$ is reached only on a worst case example. Using the algorithms D_2, WL_1, WL_λ^1, WL_λ^2 the optimality gap Δ usually does not exceed 2% and 20% for SL_λ. The knowledge of the lookahead of different sizes does not improve the solution produced by the online algorithms without lookahead. For instance, if D_2 produces the solution with $\Delta = 1.91\%$, then WL_1 decreases this value till 1.67%, WL_λ^1 till 1.48%, WL_λ^2 till 1.06%. The observation also shows that increasing the lookahead does not improve the value of Δ, *i.e.*, the number of output cycles (see Fig. 3.2.6) in most cases. This is to be expected. If the Rotastore has only two layers, not more than two objects at the same time can be distributed, therefore the knowledge of more objects (*i.e.*, $\lambda \geq 2$) has no advantages.

3.3. Summary

In this chapter we considered one real life application of BPSP: the storage system Rotastore. We demonstrated that $BPSP_1$ produced solutions quickly but very far from the optimal solution (integrality gap of almost 50%). $BPSP_2$ gives near optimal results in short time (integrality gap less than 10%). Finally, $BPSP_3$ produces optimal solutions, but the running time can become sometimes very long. Therefore, we suggest to use a combination of $BPSP_2$ and $BPSP_3$ to get solutions with integrality gaps of less than 1% in short time. For the polynomial

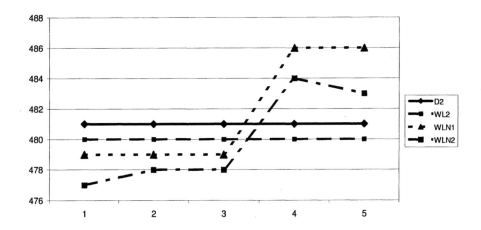

Figure 3.2.6. Solutions of the online algorithms with lower bound 472 versus size of the lookahead

case we compared all algorithms described in Section 2.5 and showed that a lookahead brings almost no further improvement, *i.e.*, the online algorithm without lookahead is already good by itself (it produces near-optimal solutions). For this reason, this online algorithm was extended to the case of 2^P layers.

Chapter 4

VEHICLE ROUTING PROBLEMS
IN HOSPITAL TRANSPORTATION. I
MODELS AND SOLUTION APPROACHES

In this chapter we focus on the VRPPDTW, adapted for hospital transportation problems. After a description of the problem (Subsection 4.1.1) and introducing some notations (Subsection 4.2.1), we suggest several approaches for solving this problem, including a mixed integer linear programming (MILP) formulation (Subsection 4.3.1), a branch-and-bound approach (Subsection 4.3.2), a column enumeration approach (Subsection 4.3.3), and heuristics methods (Section 4.4).

4.1. Problem Formulation and Solution Outline
4.1.1 Problem Description

In this subsection we give a formulation of the vehicle routing problem with pickup and delivery and time window constraints (VRPPDTW). Solutions of the problem are needed for managing the transport of patients and goods in a hospital campus by a dispatcher responsible for coordinating vehicles. The setup for the hospital-oriented VRPPDTW problem differs from the standard VRPPDTW in the following aspects:

- loads may include not only independent single items (patients or materials), but also groups of people (patients with accompanying personnel);

- target times for pickup and delivery have to be met instead of strict time windows, or the delay *w.r.t.* target times should be minimized, *respectively*;

- passengers should not spend more than, say, 30 minutes in a vehicle, *i.e.*, the time between pickup and delivery is limited;

- some items may require special transport equipment (*e.g.*, wheelchair, bed, or special vehicle);

- certain items cannot be transported simultaneously (*e.g.*, infectious patients);

- a dispatcher receives new requests or orders and needs to assign them to free vehicles or to add them to existing tours, *i.e.*, an online version of the VRPPDTW is required.

The problem is characterized as follows.

Data:

- a heterogeneous fleet of vehicles;

- a set of requests or orders;

- source and destination depots for each vehicle v, s_v and d_v, respectively;

- distances between all locations, *i.e.*, between all possible locations, including vehicle depots.

Orders:

- type of an order, *i.e.*, description of transported goods or type of transported patient to be picked up and delivered. A patient may be transported while sitting in the vehicle, lying in a bed, or sitting in a wheelchair;

- service time associated with an order at the pickup and/or at the delivery location (for instance, to load a patient lying in bed);

- target time associated with an order at pickup and/or delivery location.

Vehicles:

- time intervals during which the vehicles are available;

- type of a vehicle (*i.e.*, which type of goods/patients it can transport);

- capacity of a vehicle with respect to its type (how many goods/patients of a certain type it can transport simultaneously).

Objectives:

- to minimize the total travel time of the goods/patients (possibly some goods have to be transported very fast, *e.g.*, blood);

- to minimize the total duration of transportation for each vehicle (better exploitation of the vehicle fleet);

- to minimize possible delays at pickup and delivery locations (increase the satisfaction of patients).

Each order needs to be carried out on time (or if that is not possible, its delay should be minimized) and served by only one vehicle. Each vehicle must start from its origin depot and return to its destination depot at the end of its period of service. In this book, driven by the real world hospital case, we concentrate mostly on experiments using the third objective function. However, the other objective functions, and also combinations of them, can be considered as well.

The problem involves assigning the requests or orders to the available vehicles. The requests occur at the nodes of a transportation network and the routes are characterized by the sequence of the nodes to be visited consecutively by each vehicle. If the assignment of nodes to a vehicle has been done, both the sequence of the nodes in the routes and the arrival and departure times have to be determined. Thus, we have to solve a combined assignment, routing (sequencing) and scheduling problem:

assignment	assigning orders to vehicles
routing	putting the nodes in a route in sequence
scheduling	computing arrival and departure times, as well as considering temporal constraints

Let us comment on the temporal restrictions of the real world problem we will analyze in Chapter 5. Usually, and this also applies to the Desaulniers *et al.* formulation [25] (briefly listed and discussed in subsection 4.1.2), the temporal constraints are exact time window constraints such as (4.1.3): if the vehicle visits a node of its tour, it must arrive within a time window $[a, b]$. Should it arrive earlier than time a, the vehicle is allowed to wait. As was mentioned above our problem is different:

1 Instead of time windows only target times are to be observed, *i.e.*, desired pickup and delivery times.

2 For most orders or requests only one target time is specified and valid for either the pickup or the delivery node. As passengers should not spend more than Δ_T minutes in a vehicle, we define the target time not specified as a function of the specified one and Δ_T.

3 The target times are allowed to be met late but lateness is penalized in the objective function, *i.e.*, they are relaxable temporal constraints.

4 For pickup nodes, the vehicle must wait if it arrives earlier than the target time. In this case, the target time is to be interpreted as an earliest pickup time. Earliness, or the equivalent waiting time, is also penalized in the objective function.

Despite these features we keep calling our model a VRPPDTW (time window) model, because we do not wish to coin or add a new term to this research field. Formally, one could classify our problem as a *multi-commodity, capacitated vehicle routing problem with pickup and delivery and relaxable time window constraints*, in which for pickup nodes the upper time limit, b, is set to $+\infty$, and for delivery nodes the lower time limit, a, is set to $-\infty$.

4.1.2 Discussion of one Particular Model Formulation

In this subsection we briefly review the model presented by Desaulniers *et al.* [25], and comment on it. The model presented is, for general objective functions, a routing-scheduling model assuming that the assignment decisions have been fixed previously. Only for special objective functions, their model can solve the vehicle assignment problem as well. We adapt the notation used in [25] to that used in this book:

Identify request (we use the term *order*, synonymously) i by two nodes, i and $N+i$, corresponding, respectively, to the pickup and delivery stops of the request. It is possible that different nodes may represent the same geographical locations. Next, denote the set of pickup nodes by $\mathcal{P} = \{1, ..., N^O\}$ and the set of delivery nodes by $\mathcal{D} = \{N+1, ..., 2N^O\}$. Further, define $\mathcal{N} = \mathcal{P} \cup \mathcal{D}$. If request i consists of transporting D_i units from i to $N+i$, let $L_i = D_i$ and $L_{n+i} = -D_i$.

Let $\mathcal{V}$ be a set of N^V vehicles. Because not all vehicles can service all requests, each vehicle v may have a specific set $\mathcal{N}_v = \mathcal{P}_v \cup \mathcal{D}_v$ associated with it, where $\mathcal{N}_v \subseteq \mathcal{N}$, $\mathcal{P}_v \subseteq \mathcal{P}$, and $\mathcal{D}_v \subseteq \mathcal{D}$ are appropriate subsets of $\mathcal{N}$, $\mathcal{P}$ and $\mathcal{D}$, respectively. This subset may have been constructed within a pre-solving phase. Note that at this level nothing is said about whether the same request i is contained in several subsets. However, if the subsets $\mathcal{N}_v$, $\mathcal{P}_v$, and $\mathcal{D}_v$ have been constructed by an assignment heuristic, or by a column generation approach based on a set partitioning model, then each request i is only contained in one vehicle subset.

For each vehicle v, define now the network $\mathcal{G}_v = (\mathcal{K}_v, \mathcal{A}_v)$. Set $\mathcal{K}_v = \mathcal{N}_v \cup \{o(v), d(v)\}$ as the sets of nodes inclusive of the origin, $o(v)$, and destination, $d(v)$, depots for vehicle v, respectively. The subsets $\mathcal{A}_v$ of $\mathcal{K}_v \times \mathcal{K}_v$ contain all feasible arcs. The subsets, $\mathcal{A}_v$, may have been constructed by eliminating infeasible arcs from the full set of all arcs. The capacity of vehicle v is given by C_v, and its travel time and cost between distinct nodes $i, j \in V_k$, by T_{ijv}^D and C_{ijv}^D, respectively.

Vehicle v is assumed to leave unloaded from its origin depot at time $a_{o(v)} = b_{o(v)}$. Each admissible pickup and delivery route for this vehicle corresponds to a feasible path from $o(v)$ to $d(v)$ in network G_v, visiting each node at most once. If the vehicle visits node $n \in \mathcal{N}$, it must do so within time window $[a_n, b_n]$ when the service time T_n^S must begin. Should it arrive too early, the vehicle is allowed to wait.

The formulation involves three types of variables: binary variables δ_{ijv}, is equal to 1, if arc $(i, j) \in A_v$ is used by vehicle v, and 0 otherwise; time variables t_{nv} specifying when vehicle v starts the service at node $n \in V_k$; and variables p_{nv} giving the load of vehicle v after the service at node $n \in \mathcal{K}_v$ has been completed. The Desaulniers *et al.* formulation is as follows:

$$\min \sum_{v \in \mathcal{V}} \sum_{(i,j) \in A_v} C_{ijv}^D \delta_{ijv} \qquad (4.1.1)$$

subject to

$$\sum_{v \in \mathcal{V}} \sum_{j \in N_v \cup \{d(v)\}} \delta_{ijv} = 1 \quad , \quad \forall i \in \mathcal{P}$$

$$\sum_{j \in N_v} \delta_{ijv} - \sum_{j \in N_v} \delta_{j,N^O+i,v} = 0 \quad , \quad \forall v \in \mathcal{V},\ i \in \mathcal{P}_v,$$

$$\sum_{j \in P_v \cup \{d(v)\}} \delta_{o(v),jv} = 1 \quad , \quad \forall v \in \mathcal{V},$$

$$\sum_{i \in N_v \cup \{o(v)\}} \delta_{ijv} - \sum_{i \in N_v \cup \{d(v)\}} \delta_{jiv} = 0 \quad , \quad \forall v \in \mathcal{V},\ j \in \mathcal{N}_v,$$

$$\sum_{i \in D_v \cup \{d(v)\}} \delta_{i,d(v),v} = 1 \quad , \quad \forall v \in \mathcal{V},$$

$$\delta_{ijv}(t_{iv} + T_i^S + T_{ijv}^D - t_{jv}) \le 0 \quad , \quad \forall v \in \mathcal{V},\ (i,j) \in \mathcal{A}_v, \qquad (4.1.2)$$

$$a_n \leq t_{nv} \leq b_n \quad , \quad \forall v \in \mathcal{V}, \ n \in \mathcal{K}_v, \qquad (4.1.3)$$

$$t_{iv} + T^D_{i,N+i,v} \leq t_{N+i,v} \quad , \quad \forall v \in \mathcal{V}, \ i \in \mathcal{P}_v,$$

$$\delta_{ijv}(p_{iv} + L_j - p_{jv}) = 0 \quad , \quad \forall v \in \mathcal{V}, \ (i,j) \in \mathcal{A}_v, \qquad (4.1.4)$$

$$L_i \leq p_{iv} \leq C_v \quad , \quad \forall v \in \mathcal{V}, \ i \in \mathcal{P}_v,$$

$$0 \leq L_{N+i,v} \leq C_v - L_i \quad , \quad \forall v \in \mathcal{V}, \ N+i \in \mathcal{D}_v,$$

$$p_{o(v),v} = 0 \quad , \quad \forall v \in \mathcal{V},$$

$$\delta_{ijv} \geq 0 \quad , \quad \forall v \in \mathcal{V}, \ (i,j) \in \mathcal{A}_v,$$

$$\delta_{ijv} \in \{0,1\} \quad , \quad \forall v \in \mathcal{V}, \ (i,j) \in \mathcal{A}_v. \qquad (4.1.5)$$

Note that the objective function includes only the arc variables δ_{ijv}, but no other variables. This has some important consequences for the optimal solution of the model (4.1.1) to (4.1.5).

1 It suffices to compute the compound load variables, p_{nk}, and begin-service time variables, t_{nk}, to obtain feasible solutions. Compound means that they depend on both node, n, and vehicle, v.

2 The assignment decision, *i.e.*, the assignment of requests to vehicles, can be derived from the arc variables, δ_{ijv}. Once the assignments are known and represented by v_n, the real begin-service times, t_n, and load information, p_n, are given by $t_n = t_{nv_n} \in \{t_{nv} : v \in \mathcal{V}\}$ and $p_n = p_{nv} \in \{p_{nv} : v \in \mathcal{V}\}$.

If the objective function depends explicitly on t_n (or p_n), the variables t_n (or p_n), and δ_{ijv} need to be connected. Of course, if each request is already assigned to a specific vehicle, then the variable t_{nv} (or p_{nv}), for each node n, exists for only one vehicle v.

Note also, that due to nonlinear relations (4.1.2) and (4.1.4) the model leads to a mixed integer nonlinear programming (MINLP) problem.

4.1.3 Discussion of Existing Solution Approaches

The vehicle routing problem (VRP) is one of the basic optimization problems known to be NP-hard [86]. Variants of the VRP, such as the vehicle routing problem with time windows (VRPTW), and the vehicle routing problem with pickup and delivery (VRPPDP), the capacitated vehicle routing problem (CVRP) were thoroughly investigated during recent years. All of them are NP-hard, and thus many researchers looked for efficient solution methods solving at least instances of modest size. Early surveys on the literature of vehicle routing with time window constraints are given in [81] and [82].

Several approaches can be found in the literature:

1 **Exact approaches based on Dynamic Programming, B&B, B&C, and Column Generation algorithms.** These methods are mostly applied to the capacitated VRP. They include dynamic programming [60], branch-and-bound (B&B) algorithms ([31], [55]), branch-and-cut (B&C) algorithms ([5], [6], [8], [63], [69], [19]), and column generation algorithms ([26], [9], [48]). If time windows are included as in [26], the time window constraints are treated deterministically and are not subject to relaxation. In the early 1990's the best B&P solvers were able to solve multi-vehicles problems with up to 100 orders [26].

2 **Route construction and improvement heuristics.** The construction part of such heuristics includes sequential insertion heuristics in which further nodes are inserted into existing routes (see [67], [80]). The improvement part includes exchange heuristics [68], such as 2-opt*, 3-opt, Or-opt or CROSS-exchange with 2-opt* and Or-opt as special cases ([21], [46], [58], [81], [84], [85]), see for instance, Figures 4.1.1 and 4.1.2 for 2-opt and 3-opt, respectively.

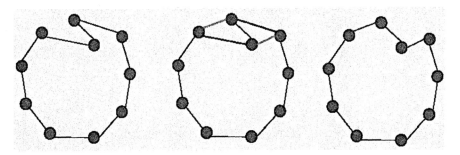

Figure 4.1.1. The exchange heuristic *2-opt*

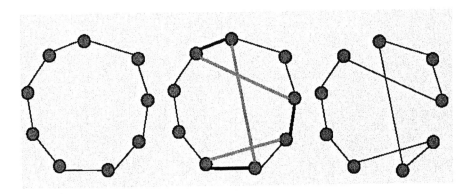

Figure 4.1.2. The exchange heuristic *3-opt*

3 **Metaheuristics.** Unlike mathematical optimization approaches, metaheuristics are based on simulating a given system or a problem and evaluating a *function of merit* (analogous to the objective function in exact optimization) [35]. Metaheuristics are not problem specific (for applications to the vehicle routing problem see, for instance, [62], [10], [16], [17], [13], [36], or [44]), and are based on generic principles and schemes that can be used to construct problem-specific heuristics: *genetic algorithms* (GA), *simulated annealing* (SA), *tabu search* (TS), and *variable neighborhood search* (VNS). All metaheuristics usually lack a proof of convergence and a proof of optimality. However they improve a given solution by performing a local search ([74], [75]), coupled with exchange mechanisms based on appropriate neighborhood relations. SA links the probability of accepting a solution that is inferior to the reference solution to a temperature-like parameter that describes the cooling of metals (see, for instance, [1], [2], [28] or [52]). TS ([20], [35], [66], [79], [84]) is a metastrategy for guiding known heuristics past the traps of local optimality. It exploits knowledge from previous solutions and thus uses a kind of memory. GA ([7], [42], [43], [65]) use a population of solutions subject to *survival of the fittest* criteria as well as mutation and recombination of positive properties in solutions. VNS ([40], [71]) explores increasingly distant neighborhoods of the current solutions and moves to a new solution only if there is an improvement.

4.1.4   Outline of Proposed Methods

In this subsection several solution approaches we developed are briefly described and implemented to solve the VRPPDTW problem of Section

4.1.1. Notice that the problem has some specific features: vehicles may have partial availability over the day, *i.e.*, very often the vehicles are available during different time intervals; vehicles have different capacities, *i.e.*, not all vehicles can carry patients that are lying down or are sitting in wheelchairs. As a consequence, orders can only be exchanged between certain vehicles fulfilling those constraints. If route improvement heuristics and VNS methods do not incorporate these features, many infeasible tours need to be evaluated.

In addition we should keep in mind the overall motivation. We wish to develop algorithms to be integrated in a software supporting the daily work of a dispatcher in a hospital, which requires solving online instances of the problem. A typical example involves, for instance, 26 orders, 16 of them already assigned to individual vehicles, and 10 of them subject to assignment, routing and scheduling to the fleet of vehicles. Thus, the algorithms should be suitable to solve small instances of the problem in short time. In order to compare the quality of the online solutions to the overall offline optimal solution over a given period, we would like to be able to solve offline problems as well. But this is not the main focus of this book, and therefore it is not necessary to solve large problems to optimality.

We consider specific features of VRPPDTW discussed in Section 4.1.1 and suggest several solution methods that include exact optimization approaches, namely mixed integer programming (Section 4.3.1), and branch-and-bound approach (Section 4.3.2), column enumeration (Section 4.3.3), and heuristic approaches (Section 4.4). The heuristics are necessary to solve VRPPDTWs whose size exceeds those that exact methods can solve. Below we give an outline of the methods we have developed and implemented:

- **Exact optimization approaches**

   1 solving a MILP model describing the intra-tour planning problem: routing and scheduling of orders within one vehicle (*single-vehicle problem*);

   2 solving a MILP model describing the inter-tour planning problem: assigning orders to vehicles, and routing and scheduling them (*multi-vehicles problem*);

   3 a branch-and-bound approach to solve the intra-tour planning problem; and

   4 a special variant of a column generation method based on complete column enumeration to solve the inter-tour planning problem.

We have adjusted all approaches to solve both online and offline scenarios.

■ **Heuristic approaches**

1 construction of initial tour sets

2 improvement of tour sets

    (a) intra-tour improvements using a sequencing heuristic

        i complete enumeration

        ii heuristic approaches, including simulated annealing (SA)

        iii solving the subproblems to optimally using the MILP model within the heuristic approach (this will only be suggested, but not implemented)

    (b) inter-tour improvements using a reassignment heuristic

        i cross-over[1] heuristics to move one order from one vehicle to another

        ii SA

    (c) combined "SH-RH-SH" tour procedure with both the sequencing and reassignment improvement strategies

The combined "sequencing-reassignment-sequencing" tour heuristic works effectively for all data we have tested. It produces a schedule for an individual vehicle or a whole fleet in reasonable time. It exploits the sequencing and reassignment improvement heuristics. For small instances the results of the heuristics are compared to the optimal solutions.

---

[1]The term *cross-over* used in this book should not be confused with its usage in the genetic algorithms community.

4.2.   General Framework

4.2.1   Notation

Prior to presenting the mathematical model and the heuristics, we introduce the required notation:

N^O : the total number of orders
N^V : the total number of vehicles
$\mathcal{V}$: the set of all vehicles, $|\mathcal{V}| = N^V$
$\mathcal{O}$: the set of all orders, $|\mathcal{O}| = N^O$
$\mathcal{P}$: the set of pickup nodes, $|\mathcal{P}| = N^O$
$\mathcal{D}$: the set of delivery nodes, $|\mathcal{D}| = N^O$
$\mathcal{N}$: the set of all pickup and delivery nodes; $\mathcal{N} = \mathcal{P} \cup \mathcal{D}$
$\mathcal{M}_v$: the set of pickup and delivery nodes extended by the depots $\{s_v, d_v\}$ for vehicle v; $\mathcal{M}_v := \mathcal{N} \cup \{s_v, d_v\}$
$\mathcal{A}$: the set of all possible arcs, (n_1, n_2), vehicle can drive along
$\mathcal{O}_v$: the set of orders served by vehicle v

To address the various objects in those sets, the following indices are used (especially, in the MILP model):

i : orders or pickup nodes of orders
j : delivery nodes
n : nodes (unspecified whether pickup or delivery node)

Nodes are usually addressed by the index n. However, if it is known that a certain node is a pickup node, then the index i is used; a similar convention applies to the use of delivery nodes j. If we need to indicate that j is a delivery node corresponding to pickup node i, then the index j_i is used. Orders and pickup nodes are used synonymously as each order uniquely induces one pickup node. Thus, the index i is used for both pickup nodes and orders.

Additionally, in the MILP model the following data are used:

$C^D_{n_1 n_2 v}$: cost for vehicle v to drive from node n_1 to node n_2
C^{PA}_{iv} : penalty cost for arriving too early at node i
C^{PB}_{iv} : penalty cost for arriving too late at node i
C^L_v : capacity of a vehicle for lying patients
C^S_v : capacity of a vehicle for sitting patients
C^W_v : capacity of a vehicle for wheelchairs
C^{LL}_{nv} : cost per lying patient in vehicle v at node n
C^{LS}_{nv} : cost per sitting patient in vehicle v at node n
C^{LW}_{nv} : cost per wheelchair in vehicle v at node n

D^S_i : demand (sitting persons) associated with order i
D^L_i : demand (lying persons) associated with order i
D^W_i : demand (persons in wheelchairs) associated with order i

L_i^S : load of sitting people at node i; $L_i^S := D_i^S$

$L_{j_i}^S$: unload of sitting people at node j_i; $L_{j_i}^S := -D_i^S$

L_i^L : load of lying people at node i; $L_i^L := D_i^L$

$L_{j_i}^L$: unload of lying people at node j_i; $L_{j_i}^L := -D_i^L$

L_i^W : load of wheelchairs at node i; $L_i^W := D_i^W$

$L_{j_i}^W$: unload of wheelchairs at node j_i; $L_{j_i}^W := -D_i^W$

T_n^S : service time for node n, *e.g.*, for loading a patient

$T_{n_1 n_2 v}^D$: driving time from node n_1 to node n_2 for vehicle v

T_n^E : earliest permissible arrival time at node n

T_n^L : latest permissible arrival time at node n

T_n^T : target time associated with node n;

            $T_n^T := T_n^E$ for pickup nodes, $T_n^T := T_n^L$ for delivery nodes

$[T_v^{VE}; T_v^{VL}]$: the time window during which vehicle v is available

T_* : current time

Δ_T : maximal time difference between delivery and pickup

R^{A_1} : maximal allowed earliness *w.r.t.* the target time at pickup nodes

R^{A_2} : maximal allowed lateness *w.r.t.* the target time at pickup nodes

R^B : maximal allowed lateness *w.r.t.* the target time at delivery nodes

Instead of time windows for pickup and delivery we use target times. Lateness *w.r.t.* these target times should be avoided or at least minimized. The earliest permissible arrival time at a pickup node, T_i^E, is identical to the desired pickup time, *i.e.*, the target time is $T_i^T = T_i^E$.

Now we formally define a tour for an arbitrary vehicle v on a set of orders, where each order is defined by two nodes: pickup node i and its corresponding delivery node j_i.

DEFINITION 4.1 *A tour $\mathcal{T}$ on a set of N^O orders is a sequence of $2N^O+2$ nodes to be visited exactly once. Each tour starts at a source depot and finishes at a destination depot. A tour is feasible if:*

1 *for every order i, its pickup node is visited before the corresponding delivery node j_i;*

2 *the load of a vehicle must never exceed its capacity, i.e., $L_i^S \le C_v^S$, $L_i^L \le C_v^L$, and $L_i^W \le C_v^W$ for any order i;*

3 *load L_i at pickup node i is the same as load L_{j_i} at the corresponding delivery node j_i, i.e., a vehicle at node j_i can only deliver the same number of persons it has picked up at node i, where L_i is L_i^S, L_i^L or L_i^W.*

As illustrated in Fig. 4.2.3 a vehicle can drive not only directly from a pickup node to a delivery node but from a pickup node to another pickup node, as well as from a delivery node to another delivery node (of course, the conditions 1 to 3 of Definition 4.1 need to be observed).

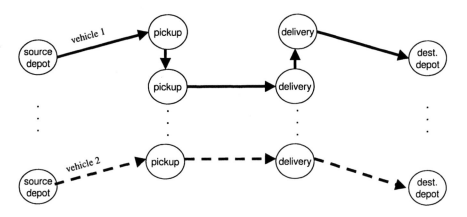

Figure 4.2.3. Possible tours for vehicles

4.2.2   Characterizing the Quality of Tours

Before discussing solution methods for VRPPDTW in detail, we first explain how the quality of a vehicle's tour is measured. The overall goal is to serve all transportation orders, *i.e.*, to pickup all patients at the requested times and transport them to their destinations. Due to the finite capacity of the vehicle fleet, however, it is not always possible to fulfill all temporal constraints (*i.e.*, delays may occur). For a specific vehicle, v, and its daily tour, we can thus compute its total lateness, S_v^D, as

$$S_v^D = \sum_{n=1}^{2|\mathcal{O}_v|} d_{vn} \quad , \quad d_{vn} := \max(0, t_n - T_n^T) \quad , \qquad (4.2.1)$$

where d_{vn} is the lateness of vehicle v in node n, and t_n and T_n^T are the actual arrival time and the target time associated with node n for vehicle v, respectively. We can also determine the average lateness per order

$$\bar{S}_v^D := S_v^D / N_v^O \quad ,$$

where N_v^O is the number of orders served by vehicle v. Additionally, we define the maximal lateness, D_v, occurring in vehicle ν. Note that $\bar{S}_v^D$ can be a more useful measure since some vehicles operate only for a few hours and thus do not serve the same amount of orders. For the whole fleet of vehicles we can further compute the total lateness, the average total lateness, and the standard deviations:

$$S := \sum_{v=1}^{N^V} \bar{S}_v^D \qquad \text{- total average lateness}$$
$$S_D := \sum_{v=1}^{N^V} D_v \qquad \text{- total maximal lateness}$$
$$\bar{S} := \frac{S}{N^V} \qquad \text{- average lateness}$$
$$\bar{S}_D := \frac{S_D}{N^V} \qquad \text{- average maximal lateness}$$
$$D_* := \max\{D_v : v \in \mathcal{V}\} \qquad \text{- the maximal lateness}$$

and

$$\sigma_S := \sqrt{\frac{1}{N^V - 1} \sum_{v=1}^{N^V} \left(\bar{S}_v^D - \bar{S}\right)^2}, \quad \sigma_D := \sqrt{\frac{1}{N^V - 1} \sum_{v=1}^{N^V} \left(D_v - \bar{S}_D\right)^2} \; .$$

Ideally, we want to have routes and schedules satisfying the condition $\bar{S} = 0$, which is equivalent to all temporal constraints being fulfilled. However, for scenarios with insufficient vehicle capacities this is not likely. Therefore several objective functions to be minimized appear promising:

$$S \quad , \quad \bar{S} \quad , \quad \sigma_S \quad , \quad S_D \quad , \quad \bar{S}_D \quad , \quad \sigma_D$$

as well as their combinations. In the context of the MILP model the objective function contains a driving cost term and penalty terms for violating the target time constraints (see 4.3.1), while in the heuristic approaches we will use, for instance, the *merit function*

$$f := S + \mu S_D$$

to measure the quality of tours, where an appropriate value of μ can be defined experimentally.

4.3.   Exact Solution Approaches

4.3.1   A Mixed Integer Programming Approach

The formulation presented by Desaulniers *et al.* [25] (briefly listed and discussed in Section 4.1.2) has been:

- modified (we have transformed their nonlinear constraints into linear ones),

- extended (including the assignment decisions for general objective functions) and

- adjusted to our needs (included special features, *e.g.*, limiting the time a passenger spends in a car, or more appropriate objective functions).

As has already been discussed on page 59 this involves the following model aspects, or decisions to be made by the solver:

- M1: assignment of orders to vehicles (not covered by Desaulniers *et al.* for general mixed-integer linear objective functions),

- M2: routing (sequencing the nodes, *i.e.*, establishing a tour for each vehicle),

- M3: scheduling (computing arrival and departure times; matching the temporal constraints), and

- M4: more objective functions.

multi-vehicles problems including the assignment problem, if solved with exact algorithms at all, are nowadays usually approached with branch-and-price algorithms including column generation (*cf.* [26], or most recently [48]). Our approach is based on a full MILP model including variables tracing the assignment decisions. The model developed below can also support several returns of a vehicle to its depot, *i.e.*, when a vehicle has several service periods during the day.

As the problem is NP-hard we expect that only small size problems can be solved, or just single-vehicle problems. Nevertheless optimal solutions help us to check the quality of the tours our heuristics generate. Depending on vehicle availability, we are able to find an optimal solution with up to 15 orders and a few vehicles, or up to 25 orders with only one vehicle (only M2 and M3) within a few minutes using the commercial MILP-solver Xpress-MP [41]; see the examples described in Chapter 5. The MILP-model can be used to calibrate certain strategies such as 2-opt or similar techniques in the heuristic approach.

4.3.1.1 Data, Indices and Variables

The routing and scheduling decisions are represented by the binary flow variables $\delta_{n_1 n_2 v}$,

$$\delta_{n_1 n_2 v} := \begin{cases} 1 \text{ , if vehicle } v \text{ drives from } n_1 \text{ to node } n_2 & \forall (n_1, n_2) \in \mathcal{A} \\ 0 \text{ , otherwise} & \forall v \in \mathcal{V} \end{cases}$$

and the assignment binary variables α_{nv},

$$\alpha_{nv} := \begin{cases} 1 \text{ , if vehicle } v \text{ serves node } n & \forall n \in \mathcal{N} \\ 0 \text{ , otherwise} & \forall v \in \mathcal{V} \end{cases}$$

As orders and pickup nodes are represented by the same index, $\alpha_{iv} = 1$ implies that order i is assigned to vehicle v. If a single-vehicle problem or a problem with known assignment is to be solved, the α_{iv} variables are just set to one. Here, and whenever the context leads to a unique interpretation, we use the abbreviated canonical indexing, *e.g.*,

$$\forall\{nv\} := \{\forall n, v : n \in \mathcal{N}, \quad v \in \mathcal{V}\} \quad .$$

Furthermore, the model contains the following non-negative continuous variables:

p_{nv}^S : number of sitting people in vehicle v when leaving node n
p_{nv}^L : number of lying people in vehicle v when leaving node n
p_{nv}^W : number of wheelchairs in vehicle v when leaving node n

t_{nv}^A : arrival time of vehicle v at node n
t_{nv}^D : departure time of vehicle v from node n

$r_{iv}^{A_1}$: earliness *w.r.t.* the target time at pickup node i of vehicle v
$r_{iv}^{A_2}$: lateness *w.r.t.* the target time at pickup node i of vehicle v
r_{jv}^{B} : lateness *w.r.t.* the target time at delivery node j of vehicle v

4.3.1.2   The Objective Function

The objective function, z,

$$z := c^{\mathrm{T}} + c^L + c^P + c^N \quad , \quad c^{\mathrm{T}} := c^D + c^{A_1} + c^{A_2} + c^B \qquad (4.3.1)$$

to be minimized contains several non-negative components:

1 driving costs, c^D;

2 penalty costs, c^{A_1}, c^{A_2}, and c^B associated with the relaxation variables used to model the target time restrictions as soft constraints;

3 capacity load costs c^L;

4 penalty costs, c^P, for deviations from given target tours, $\delta_{n_1 n_2 v}^*$. This can be useful in the online case if the aim is to re-schedule the tours and keep the re-scheduled tours as close as possible to the existing ones;

5 a term, c^N, added for numerical reasons giving preference to solutions in which vehicles arrive at the delivery node no later than necessary.

The individual components are defined as follows. The driving costs are given by

$$c^D := \sum_{v=1}^{N^V} \sum_{(n_1, n_2) \in \mathcal{A}} C_{n_1 n_2 v}^D \delta_{n_1 n_2 v} \quad . \qquad (4.3.2)$$

If all values $C^D_{n_1 n_2 v}$ are set to the same constant value, *e.g.*, $C^D_{n_1 n_2 v} = 1$, then c^D just gives the number of arcs in the graph. The model considers the three penalty terms related to the time window or target times: respectively, constraints (4.3.15)-(4.3.17):

early arrivals at pickup nodes	$c^{A_1} := \sum_{v=1}^{N^V} \sum_{i \in \mathcal{P}} C^{PA}_{iv} r^{A_1}_{iv}$
late arrivals at pickup nodes	$c^{A_2} := \sum_{v=1}^{N^V} \sum_{i \in \mathcal{P}} C^{PB}_{iv} r^{A_2}_{iv}$
late arrivals at delivery nodes	$c^B := \sum_{v=1}^{N^V} \sum_{j \in \mathcal{D}} C^{PB}_{jv} r^B_{jv}$.

There might be real costs or penalty costs proportional to the load of a car, *i.e.*,

$$c^L := \sum_{i \in \mathcal{P}} \sum_{v=1}^{N^V} C^{LL}_{iv} p^L_{iv} + \sum_{i \in \mathcal{P}} \sum_{v=1}^{N^V} C^{LS}_{iv} p^S_{iv} + \sum_{i \in \mathcal{P}} \sum_{v=1}^{N^V} C^{LW}_{iv} p^W_{iv} \quad . \quad (4.3.3)$$

This term had been added to model the situation that a vehicle's capacity (for instance, lying capacity) is two, but that it is somewhat inconvenient to carry two lying passengers. This situation is slightly avoided by assigning non-zero values to C^{LL}_{iv}, C^{LS}_{iv} or C^{LW}_{iv}.

Penalizing the deviation from target tours can be realized by adding the following term to the objective function:

$$c^P := \sum_{v=1}^{N^V} \sum_{(n_1, n_2) \in \mathcal{A}} C^P_{n_1 n_2 v} \left| \delta_{n_1 n_2 v} - \delta^*_{n_1 n_2 v} \right| \quad ,$$

where $C^P_{n_1 n_2 v}$ is a penalty cost for deviation from the target tour for vehicle v supposed to drive from node n_1 to n_2.

The term, c^N,

$$c^N := \varepsilon \sum_{j \in \mathcal{D}} \sum_{v=1}^{N^V} t^A_{jv} \quad , \quad \varepsilon := 0.00001$$

is only added for numerical reasons. As the arrival times are coupled to the departure times by equalities involving the driving time, this is not necessary from an algebraic point of view. However, it improves the efficiency of the branch-and-bound process (helps to produce feasible solutions somewhat faster).

4.3.1.3 The Constraints

Network Flow Constraints. The following constraints apply:

Every pickup node, i, is served exactly once and by one vehicle, *i.e.*,

$$\sum_{v=1}^{N^V} \sum_{n \in \mathcal{N}} \delta_{inv} = 1 \quad , \quad \forall i \in \mathcal{P} \quad . \tag{4.3.4}$$

If a pickup node, i, is visited by a given vehicle, then the corresponding delivery node[2], j_i, must be visited as well, *i.e.*,

$$\sum_{n \in \mathcal{N}} \delta_{inv} - \sum_{n \in \mathcal{N}} \delta_{n j_i v} = 0 \quad , \quad \forall i \in \mathcal{P} \quad , \quad \forall v \in \mathcal{V} \quad . \tag{4.3.5}$$

Each delivery node is also served exactly once and by the same vehicle that serves the corresponding pickup node. Therefore, the equalities (4.3.4) and (4.3.5) ensure that each order is served once and by the same vehicle.

Each vehicle starts from its source depot. It can drive directly from the source depot to the destination depot, meaning that it did not serve any order, *i.e.*,

$$\sum_{n \in \mathcal{P} \cup \{d_v\}} \delta_{s_v n v} = 1 \quad , \quad \forall v \in \mathcal{V} \quad . \tag{4.3.6}$$

Each vehicle ends its route at its destination depot, *i.e.*,

$$\sum_{n \in \mathcal{D} \cup \{s_v\}} \delta_{n d_v v} = 1 \quad , \quad \forall v \in \mathcal{V} \quad . \tag{4.3.7}$$

Every destination node has a source node (flow conservation at each node is guaranteed), *i.e.*,

$$\sum_{n_1 \in \mathcal{M}_v \setminus \{d_v\}} \delta_{n_1 n v} - \sum_{n_1 \in \mathcal{M}_v \setminus \{s_v\}} \delta_{n n_1 v} = 0 \quad , \quad \begin{matrix} \forall v \in \mathcal{V}, \\ \forall n \in \mathcal{N}, \quad n \neq n_1. \end{matrix}$$

$$\tag{4.3.8}$$

Temporal Constraints. The temporal constraints basically involve the arrival and departure times, t_{nv}^A and t_{nv}^D. The model considers constraints

1 relating the departure time from a source node with the arrival time at a destination node,

2 restricting the time a passenger spends in the vehicle,

---

[2]If the orders or pickup nodes, respectively, are indexed by numerical indices $1, \ldots, N^O$, then the delivery node j_i, corresponding to the delivery i, is given by $j_i := i + N^O$. In other words, $\mathcal{P} = \{1, ..., N^O\}$ and $\mathcal{D} = \{N^O + 1, ..., 2N^O\}$.

3 expressing either the time window restrictions or the target time restrictions.

Further, we explain each group of constraints 1-3 in detail:

1) The departure time, $t^D_{n_1 v}$, at node n_1 and the arrival time, $t^A_{n_2 v}$, at node n_2 are connected by the driving time, $T^D_{n_1 n_2 v}$, *i.e.*,

$$\delta_{n_1 n_2 v}\left(t^D_{n_1 v} + T^D_{n_1 n_2 v} - t^A_{n_2 v}\right) = 0, \quad \forall (n_1, n_2) \in \mathcal{A} \quad, \quad \forall v \in \mathcal{V} \ . \tag{4.3.9}$$

If $\delta_{n_1 n_2 v} = 0$ then (4.3.9) holds. If $\delta_{n_1 n_2 v} = 1$, then $t^D_{n_1 v} + T_{n_1 n_2 v} - t^A_{n_2 v} = 0$ must be fulfilled, *i.e.*, the arrival time at the destination node n_2 is equal to the departure time from the origin node n_1 plus the time required to drive from node n_1 to node n_2. If node n is not served by vehicle v, then by definition $t^D_{nv} = t^A_{nv} = 0$. The nonlinear complementarity constraint (4.3.9) is discussed on page 79 and replaced by a system of linear inequalities.

To tighten the model a set of valid inequalities is added. Observe that the earliest departure time, t^D_{iv}, from pickup node, i, satisfies the inequalities

$$t^D_{iv} \geq \left(T^E_i + T^S_i\right)\alpha_{iv} \quad, \quad \forall i \in \mathcal{P} \quad, \quad \forall v \in \mathcal{V} \ . \tag{4.3.10}$$

This relation can be tightened by the inequalities

$$t^D_{iv} \geq t^A_{iv} + T^S_i \alpha_{iv} \quad, \quad \forall i \in \mathcal{P} \quad, \quad \forall v \in \mathcal{V} \ . \tag{4.3.11}$$

Note that the earliest arriving time, T^E_i, sets a lower limit for departure (the patient might not be ready). Departure is delayed by the service time, T^S_i. As a vehicle may arrive later than T^E_i, condition (4.3.11) might be tighter than (4.3.10).

Observe that the earliest departure time from a delivery node j must satisfy the constraint

$$t^D_{jv} \geq t^A_{jv} + T^S_j \alpha_{jv} \quad, \quad \forall j \in \mathcal{D}, \quad \forall v \in \mathcal{V} \ . \tag{4.3.12}$$

Note that we list (4.3.10) and (4.3.12) and thus distinguish between pickup and delivery nodes because there is a difference for pre-assigned orders in the online case (in the online case we do not consider constraints for pickup nodes because the patients are already in vehicles). Furthermore, it should be observed that each delivery node is visited later than the associated pickup node,

$$t^D_{iv} + T^D_{ij_i v}\alpha_{iv} \leq t^A_{j_i v} \quad, \quad \forall i \in \mathcal{P} \quad, \quad \forall v \in \mathcal{V} \ . \tag{4.3.13}$$

The inequality (4.3.13) ensures that the triangular inequality expressing that the direct connection is the shortest one is possible.

2) A patient should not be delivered later than Δ_T hours after pickup

$$\alpha_{iv}(t^A_{j_iv} - t^A_{iv} - T^S_i - \Delta_T) \leq 0 \quad . \tag{4.3.14}$$

3) Time window constraints usually take the form

$$T^E_n \leq t^A_{nv} \leq T^L_n \quad , \quad \forall n \in \mathcal{N} \quad ,$$

where T^E_n and T^L_n are given earliest and latest times. However, it may be unwise to apply the time windows as hard constraints because the problem might become infeasible.

In the present problem it is more appropriate to use target time restrictions and to treat them as soft constraints. Instead of the time window limits, T^E_n and T^L_n, only the target times $T^T_n := T^E_n$ or $T^T_n := T^L_n$ are used. The earliness and lateness *w.r.t.* the target times of the pickup nodes follow as:

$$T^E_i \alpha_{iv} - r^{A_1}_{iv} \leq t^A_{iv} \quad , \quad \forall i \in \mathcal{P} \quad , \quad \forall v \in \mathcal{V} \quad , \tag{4.3.15}$$

and

$$T^L_i \alpha_{iv} + r^{A_2}_{iv} \geq t^A_{iv} \quad , \quad \forall i \in \mathcal{P} \quad , \quad \forall v \in \mathcal{V} \quad . \tag{4.3.16}$$

Likewise, the lateness *w.r.t.* the target times of arriving at delivery nodes is given by

$$t^A_{jv} \leq T^L_j \alpha_{jv} + r^B_{jv} \quad , \quad \forall j \in \mathcal{D} \quad , \quad \forall v \in \mathcal{V} \quad . \tag{4.3.17}$$

Note that the inequalities (4.3.15)-(4.3.17) contain the assignment variables α_{iv}. If vehicle v really serves order i (*i.e.*, drives to pickup node i or to the corresponding delivery node j_i), then $\alpha_{iv} = 1$ leads to the inequalities

$$
\begin{aligned}
T^E_i - r^{A_1}_{iv} &\leq t^A_{iv} \quad , \quad \forall i \in \mathcal{P} \quad , \quad \forall v \in \mathcal{V} \quad , \\
T^L_i + r^{A_2}_{iv} &\geq t^A_{iv} \quad , \quad \forall i \in \mathcal{P} \quad , \quad \forall v \in \mathcal{V} \quad , \\
t^A_{jv} &\leq T^L_j + r^B_{jv} \quad , \quad \forall j \in \mathcal{D} \quad , \quad \forall v \in \mathcal{V} \quad .
\end{aligned}
$$

Otherwise, if $\alpha_{iv} = 0$, (4.3.15)-(4.3.17) reduce to

$$
\begin{aligned}
-r^{A_1}_{iv} &\leq t^A_{iv} \quad , \quad \forall i \in \mathcal{P} \quad , \quad \forall v \in \mathcal{V} \quad , \\
r^{A_2}_{iv} &\geq t^A_{iv} \quad , \quad \forall i \in \mathcal{P} \quad , \quad \forall v \in \mathcal{V} \quad , \\
t^A_{jv} &\leq r^B_{jv} \quad , \quad \forall j \in \mathcal{D} \quad , \quad \forall v \in \mathcal{V} \quad .
\end{aligned}
$$

Note that in this case $t^A_{iv} = t^A_{jv} = r^{A_1}_{iv} = r^{A_2}_{iv} = r^B_{jv} = 0$ are feasible values.

As the relaxation variables $r_{iv}^{A_1}$, $r_{iv}^{A_2}$ and r_{jv}^{B} appear in the objective function as penalty terms, the target time restrictions (4.3.15)-(4.3.17) can be interpreted as soft constraints that penalize deviations from the target times. The target time restrictions can be transformed into hard constraints by putting upper bounds on the relaxation variables. By coupling the relaxation variables with the assignment variables α_{iv} the model is tightened and the performance of the solution algorithm is improved:

$$r_{iv}^{A_1} \leq R^{A_1} \alpha_{iv} \quad , \quad \forall i \in \mathcal{P} \quad , \quad \forall v \in \mathcal{V} \quad , \tag{4.3.18}$$

$$r_{iv}^{A_2} \leq R^{A_2} \alpha_{iv} \quad , \quad \forall i \in \mathcal{P} \quad , \quad \forall v \in \mathcal{V} \quad , \tag{4.3.19}$$

and

$$r_{jv}^{B} \leq R^{B} \alpha_{jv} \quad , \quad \forall j \in \mathcal{D} \quad , \quad \forall v \in \mathcal{V} \quad , \tag{4.3.20}$$

where R^{A_1}, R^{A_2} and R^{B} denote the maximum allowed deviations from target times, *e.g.*, 0.5 hours. However, if the maximal value is too small, the problem might become infeasible.

Note that in the current case, there is no real reason why a vehicle should appear too early at a pickup node because it cannot leave the pickup node before its target time. But because the equality (4.3.9) relates the departure time from a source node and the arrival time at a destination node, non-zero values of $r_{iv}^{A_1}$ may be interpreted as waiting time at the destination node.

Capacity Constraints. The capacity constraints balance and track the load of the vehicles, *i.e.*, they relate the load after leaving a node to the load state before arriving at the node: for sitting passengers,

$$\delta_{n_1 n_2 v} \left(p_{n_1 v}^{S} + L_{n_2}^{S} - p_{n_2 v}^{S} \right) = 0 \quad , \quad \begin{matrix} \forall v \in \mathcal{V}, \\ \forall (n_1 n_2) \in \mathcal{A}, \end{matrix} \tag{4.3.21}$$

for lying passengers,

$$\delta_{n_1 n_2 v} \left(p_{n_1 v}^{L} + L_{n_2}^{L} - p_{n_2 v}^{L} \right) = 0 \quad , \quad \begin{matrix} \forall v \in \mathcal{V}, \\ \forall (n_1 n_2) \in \mathcal{A}, \end{matrix} \tag{4.3.22}$$

and for wheelchair passengers

$$\delta_{n_1 n_2 v} \left(p_{n_1 v}^{W} + L_{n_2}^{W} - p_{n_2 v}^{W} \right) = 0 \quad . \quad \begin{matrix} \forall v \in \mathcal{V}, \\ \forall (n_1 n_2) \in \mathcal{A} \end{matrix} \cdot \tag{4.3.23}$$

If $\delta_{n_1 n_2 v} = 1$, then the terms in parentheses become active, for instance, $p_{n_2 v}^{L} = p_{n_1 v}^{L} + L_{n_2}^{L}$.

The capacity constraints to be observed by all vehicles at the pickup nodes are given by

$$L_i^{S} \alpha_{iv} \leq p_{iv}^{S} \leq C_v^{S} \quad , \quad \forall i \in \mathcal{P} \quad , \quad \forall v \in \mathcal{V} \quad , \tag{4.3.24}$$

$$L_i^L \alpha_{iv} \leq p_{iv}^L \leq C_v^L \quad , \quad \forall i \in \mathcal{P} \quad , \quad \forall v \in \mathcal{V} \quad , \tag{4.3.25}$$

and

$$L_i^W \alpha_{iv} \leq p_{iv}^W \leq C_v^W \quad , \quad \forall i \in \mathcal{P} \quad , \quad \forall v \in \mathcal{V} \quad . \tag{4.3.26}$$

Likewise, the capacity constraints to be observed by all vehicles at the delivery nodes, are

$$0 \leq p_{jv}^S + L_j^S \alpha_{jv} \leq C_v^S \quad , \quad \forall j \in \mathcal{D} \quad , \quad \forall v \in \mathcal{V} \quad , \tag{4.3.27}$$

$$0 \leq p_{jv}^L + L_j^L \alpha_{jv} \leq C_v^L \quad , \quad \forall j \in \mathcal{D} \quad , \quad \forall v \in \mathcal{V} \quad , \tag{4.3.28}$$

and

$$0 \leq p_{jv}^W + L_j^W \alpha_{jv} \leq C_v^W \quad , \quad \forall j \in \mathcal{D} \quad , \quad \forall v \in \mathcal{V} \quad . \tag{4.3.29}$$

The vehicle loads at the depots are by definition

$$p_{s_v v}^S = p_{s_v v}^L = p_{s_v v}^W = 0 \quad , \quad \forall v \in \mathcal{V} \quad .$$

Several Tour Constraints. The model developed in [25] allows each vehicle to have only one tour (driving from the source depot to the destination depot, which in our case is the same as the source depot location). To allow for several tours (leaving the source depot several times) we assume that in the considered time interval (one day) there are maximum N^T tours. In this case all variables, described in Subsection 4.3.1.1 should also depend on the tour index t, for instance:

$$\delta_{n_1 n_2 vt} := \begin{cases} 1 \text{ , if vehicle } vt \text{ drives from } n_1 \text{ to node } n_2 \\ 0 \text{ , else} \end{cases} \quad \begin{aligned} &\forall (n_1, n_2) \in \mathcal{A} \\ &\forall v \in \mathcal{V} \\ &\forall t = 1, \ldots, N^T \end{aligned}$$

and

$$\alpha_{nvt} := \begin{cases} 1 \text{ , if vehicle } vt \text{ serves node } n \\ 0 \text{ , else} \end{cases} \quad \begin{aligned} &\forall n \in \mathcal{N} \\ &\forall v \in \mathcal{V} \\ &\forall t = 1, \ldots, N^T \end{aligned} \quad .$$

Our modeling approach is as follows. The departure time from the source depot s_v for a subsequent tour t cannot be earlier than the return time $t_{d_v, v, t-1}^A$ of the same vehicle from the previous tour $t - 1$. Tours can be connected by

$$t_{nvt}^A \geq t_{d_v vt-1}^A + T_{s_v nvt}^D \quad , \quad \begin{aligned} &\forall n \in \mathcal{N} \\ &\forall v \in \mathcal{V} \\ &\forall t = 2, \ldots, N^T \end{aligned} \quad .$$

No time window constraints are considered for depot nodes s_v and d_v.

Assignment Constraints. The assignment variable α_{iv}, besides representing the assignment decision, is useful to relax the time window constraints or, in the online case, to implement pre-given assignments of orders to vehicles. It is related to the flow variables by the equality

$$\alpha_{iv} = \sum_{n \in \mathcal{M}_v \setminus \{d_v\}} \delta_{niv} \quad , \quad \forall i \in \mathcal{P} \quad , \quad \forall v \in \mathcal{V} \quad , \tag{4.3.30}$$

which expresses that "if order i is assigned to vehicle v, then there exists an arc with destination i". Similar conditions hold for the delivery nodes j

$$\alpha_{jv} = \sum_{n \in \mathcal{M}_v \setminus \{s_v\}} \delta_{njv} \quad , \quad \forall j \in \mathcal{D} \quad , \quad \forall v \in \mathcal{V} \quad . \tag{4.3.31}$$

Furthermore, if vehicle v serves order i (or pickup node i), then it also serves the corresponding delivery node, j_i, *i.e.*,

$$\alpha_{iv} = \alpha_{j_i v} \quad , \quad \forall i \in \mathcal{P} \quad , \quad \forall v \in \mathcal{V} \quad . \tag{4.3.32}$$

Each order can be served by one vehicle v, *i.e.*,

$$\sum_{v=1}^{N^V} \alpha_{iv} = 1 \quad , \quad \forall i \in \mathcal{P} \quad . \tag{4.3.33}$$

If vehicle v drives to pickup node i then it also drives to its associated delivery node j_i

$$\alpha_{iv} = \sum_{n \in \mathcal{N}} \delta_{nj_i v} \quad , \quad \forall i \in \mathcal{P} \quad , \quad \forall v \in \mathcal{V} \quad . \tag{4.3.34}$$

If this model is used for routing and scheduling only, *i.e.*, for only one vehicle, then, $|\mathcal{V}| = 1$ and obviously

$$\alpha_{iv} = 1 \quad , \quad \forall i \in \mathcal{P} \quad , \quad \forall v \in \mathcal{V} \quad .$$

4.3.1.4   Equivalent Mixed Integer Linear Formulations

As it is quite complicated to solve nonlinear models, it is worthwhile to transform the nonlinear constraints into linear ones for the model described at the beginning of the current section. The first group of nonlinear constraints (4.3.9) implies that if $\delta_{n_1 n_2 v} = 1$, then $t^D_{n_1 v} + T^D_{n_1 n_2 v} - t^A_{n_2 v} = 0$. This is represented by the following linear inequality

$$t^D_{n_1 v} + T^D_{n_1 n_2 v} \le t^A_{n_2 v} + (1 - \delta_{n_1 n_2 v}) M_{n_1 n_2 v}, \quad \forall (n_1, n_2) \in \mathcal{A} \quad , \quad \forall v \in \mathcal{V}$$

where $M_{n_1 n_2 v}$ is a sufficiently large number, *e.g.*,

$$M_{n_1 n_2 v} := \max\{T^L_{n_1} + T^D_{n_1 n_2 v} - T^E_{n_2}, 0\} \quad .$$

So, the case $\delta_{n_1 n_2 v} = 1$ leads to $t^D_{n_1 v} + T^D_{n_1 n_2 v} \leq t^A_{n_2 v}$ and, together with (4.3.12), to $t^D_{n_1 v} + T^D_{n_1 n_2 v} = t^A_{n_2 v}$.

The nonlinear constraints (4.3.14) can be rewritten as follows:

$$t^A_{jiv} \leq t^A_{iv} + T^S_i + \Delta T + M^D - M^D \alpha_{iv} \quad , \quad \forall i \in \mathcal{P} \quad , \quad \forall v \in \mathcal{V} \quad , \quad (4.3.35)$$

where M^D is a sufficiently large constant, e.g., $M^D = 4$.

Finally, if we set

$$M^S_{n_1 n_2 v} := \max\{C^S_v\}, \quad M^L_{n_1 n_2 v} := \max\{C^L_v\}, \quad M^W_{n_1 n_2 v} := \max\{C^W_v\}, \quad (4.3.36)$$

the equations (4.3.21) can be linearized. If $\delta_{n_1 n_2 v} = 0$, then $p^S_{n_1 v} + L^S_{n_1} - p^S_{n_2 v} \leq M^S_{n_1 n_2 v}$ always holds, in case $\delta_{n_1 n_2 v} = 1$, $p^S_{n_1 v} + L^S_{n_1} - p^S_{n_2 v} = 0$, because all values of $p^S_{n_1 v}$, $L^S_{n_1}$, $p^S_{n_2 v}$ are nonnegative. So, the first of the three constraints (4.3.21) is transformed into

$$p^S_{iv} + L^S_i - p^S_{jv} \leq M^S_{ijv}(1 - \delta_{ijv}) \quad , \quad \forall (i,j) \in \mathcal{A} \quad , \quad \forall v \in \mathcal{V}.$$

The two other groups of constraints (4.3.22 and 4.3.23) are subject to the same transformation.

4.3.1.5   Online Version of the MILP Model

In the online version, the MILP model supports re-planning of existing tours or adding new orders to existing tours. At the current time T_*, all vehicles may be positioned at known location. After re-planning, each vehicle continues from where it served the last order. It may also happen that a vehicle is in transit with loaded patients/goods when re-planning occurs.

When the MILP model is used for online situations there is usually only a small set of orders involved; see, for instance, the example presented in Chapter 5. Thus, the problem can be approached fully with all its components:

- assigning orders to vehicles,

- routing (sequencing),

- scheduling (matching the temporal constraints).

Let $\mathcal{O}^P$ denote the set of pre-assigned orders, that is orders for which the passengers have already been loaded to a vehicle and still are in the car at time point T_*. If the patients of order $i \in \mathcal{O}^P$ are in vehicle v_i, then $\mathcal{O}^P_v$ denotes the orders pre-assigned to vehicle v. For order $i \in \mathcal{O}^P$ no assignment problem needs to be solved, because the assignment variables are just set to

$$\alpha_{iv_i} = 1, \quad \forall i \in \mathcal{O}^P \quad .$$

The temporal constraints related to pickup nodes of pre-assigned orders, $i \in \mathcal{O}^P$, and, hence, the variables $r_{iv}^{A_1}$ and $r_{iv}^{A_2}$ are neglected and set to zero because the patients have already been loaded to the vehicle. However, for such orders, the actual arrival times, t_{iv}^A, are computed using the given information on the time the passenger spends already in a vehicle. This allows to apply (4.3.35) to limit the total time a passenger spends in a vehicle.

The part of a tour related to picking up pre-assigned patients is seen as a fixed part of the tour, *i.e.*, it is known a priori. The pre-assignment information is used to fix the corresponding binary variables α_{iv}. From this, consistent initial conditions regarding the load variables generated. The real cases usually involve only sets $\mathcal{O}_v^P$ with a small number of orders, usually only one, two or three, *i.e.*, $|\mathcal{O}_v^P| \leq 3$.

The case $|\mathcal{O}_v^P| = 1$ allows to fix the binary variables $\delta_{s_v iv}$

$$\delta_{s_v iv} = 1, \quad i \in \mathcal{O}_v^P \quad , \quad \forall v \in \mathcal{V} \tag{4.3.37}$$

expressing that the vehicle drives from its source location, s_v, to the pickup node of order i. Note that in the offline version s_v is identical to the source depot while in the online version it can be any location. The load variables can also be fixed, *i.e.*,

$$p_{iv}^S = L_i^S, \quad i \in \mathcal{O}_v^P \quad , \quad \forall v \in \mathcal{V}$$

$$p_{iv}^L = L_i^L, \quad i \in \mathcal{O}_v^P \quad , \quad \forall v \in \mathcal{V}$$

and

$$p_{iv}^W = L_i^W, \quad i \in \mathcal{O}_v^P \quad , \quad \forall v \in \mathcal{V} \quad .$$

The case $|\mathcal{O}_v^P| = 2$ with pre-assigned orders i_{1v} and i_{2v} is more complicated because the car might have first driven to the location associated with order i_{1v}, and then to i_{2v}, or vice versa. Therefore, the constraints are:

$$\sum_{i \in \mathcal{O}_v^P} \delta_{s_v iv} = 1, \quad \forall v \in \mathcal{V} \quad ,$$

$$\delta_{s_v i_{1v} v} + \delta_{i_{1v} i_{2v} v} + \delta_{s_v i_{2v} v} + \delta_{i_{2v} i_{1v} v} = 2 \quad , \quad \forall v \in \mathcal{V} \quad ,$$

$$p_{i_{1v} v}^S = L_{i_{1v} v}^S + L_{i_{2v} v}^S \delta_{s_v i_{2v} v} \quad , \quad \forall v \in \mathcal{V} \quad ,$$

$$p_{i_{2v} v}^S = L_{i_{2v} v}^S + L_{i_{1v} v}^S \delta_{s_v i_{1v} v} \quad , \quad \forall v \in \mathcal{V} \quad ,$$

$$p_{i_{1v} v}^L = L_{i_{1v} v}^L + L_{i_{2v} v}^L \delta_{s_v i_{2v} v} \quad , \quad \forall v \in \mathcal{V} \quad ,$$

$$p^L_{i_2 v v} = L^L_{i_2 v v} + L^L_{i_1 v v} \delta_{s_v i_1 v v} \quad , \quad \forall v \in \mathcal{V} \quad ,$$

and

$$p^W_{i_1 v v} = L^W_{i_1 v v} + L^W_{i_2 v v} \delta_{s_v i_2 v v} \quad , \quad \forall v \in \mathcal{V} \quad ,$$

$$p^W_{i_2 v v} = L^W_{i_2 v v} + L^W_{i_1 v v} \delta_{s_v i_1 v v} \quad , \quad \forall v \in \mathcal{V} \quad . \tag{4.3.38}$$

Note that alternatively to the constraints (4.3.37) to (4.3.38) the information about the pre-assigned orders could also be exploited to eliminate the associated variables completely.

The current time, T_*, is used to tighten the model. If the target time, T^T_n, of a node n representing a pickup or delivery node associated with an order $i \notin \mathcal{O}^P$ is earlier than T_*, for all vehicle $v \in \mathcal{V}$ the following bounds can be added to the model:

$$r^{A_2}_{iv} \geq T_* - T^T_i, \quad \text{if } i \text{ is the pickup node associated with order } i \notin \mathcal{O}^P \quad ,$$

and

$$r^B_{j_i v} \geq T_* - T^T_{j_i}, \quad \text{if } j_i \text{ is the delivery node associated with order } i \notin \mathcal{O}^P$$

The case $|\mathcal{O}^P_v| = 3$ (or more) leads to similar constraints.

4.3.1.6   Comments on the Size and the Structure of the MILP Problem

The problem formulated above is NP-hard since the standard VRP is a specific case of it and the latter is NP-hard [86]. Even for small cases one experiences an exponential growth of the number of nodes of the branch-and-bound tree. This can be understood as follows. If we consider only the assignment part of the problem with $n = N^V$ vehicles and $r = N^O$ orders, there exist n^r possible combinations of assigning these orders to vehicles. This problem is equivalent to assigning r distinguishable objects into n cells.

THEOREM 4.2 *There exists $N(n,r) = n^r$ combinations of how r distinguishable requests can be assigned to n vehicles.*

Proof.   The proof is based on complete induction. Obviously, the statement is true for $n = 1$ and any number of requests, $r \in \mathbb{N}$; there is only $1 = 1^r$ way to achieve this.

a) Let us first do the induction step from r to $r + 1$, and derive the number $N(n, r + 1)$ from the known number $N(n, r)$. If we add one request, then we take the set of all n^r combinations obtained for r requests to n vehicles. Obviously, the new request can be assigned to any of these

n vehicles, which increases the number of elements the given set of combinations by a factor of n, *i.e.*, we have $N(n, r+1) = nN(r) = nn^r = n^{r+1}$, (q.e.d.).

b) Now we focus on the induction step for the same number of requests. Let us increase the number of vehicles by one and show that $N(n+1, r) = (n+1)^r$. The new vehicle, $n+1$, can have any number, b, of requests between 0 and r, leaving $r - b$ requests for the original n vehicles. Note that there exists $\binom{r}{b}$ combinations of how to extract b requests from a set of r requests. Therefore, the number of combinations $N(n+1, r)$ becomes

$$N(n+1, r) = \sum_{b=0}^{r} \binom{r}{b} N(n, r-b) = \sum_{b=0}^{r} \binom{r}{b} n^{r-b} = (n+1)^r \quad ,$$

where we have exploited our induction knowledge about $N(n, r-b) = n^{r-b}$ and the well known relation $(x+y)^r = \sum_{b=0}^{r} \binom{r}{b} x^{r-b} y^b$ (q.e.d). ∎

The binary variables, α_{iv}, carry the assignment of orders to vehicles but they hide the fact there exists an enormous number of combinations. An underlying assumption of the proof is that both the requests (orders) and the vehicles are distinguishable. This is strictly speaking true in our case, but the vehicles may be almost alike in terms of the capacities but may have different working periods. The orders are usually distinguishable, but they are often very much alike regarding the time windows or target times which differ only little. Thus the assignment problem bears a lot of symmetry or nearby symmetry when objective functions are used which measure the delays of arrival times. As a result, the branch-and-bound tree becomes huge.

Let us now investigate the complexity of the routing and scheduling problem (intra-tour problem). Here we have to put $m \leq r$ (distinguishable) objects into the right sequence. This gives at most $m!$ permutations. In our case for each vehicle v, $m = 2N_v^O$, with N_v^O denoting the number of orders served by vehicle v. As each pickup node precedes its delivery node, there are at most

$$N_v^R := \frac{(2N_v^O)!}{2^{N_v^O}} \tag{4.3.39}$$

combinations or routes to be tested for each vehicle. Note that the denominator cancels out a factor of two for each pickup-delivery node pair. For certain kinds of objective functions we have, in Section 4.3.2, formulated a branch-and-bound approach to solve the intra-tour optimization problem.

Despite the hardness of the problem, for small instances the routing and scheduling problem, we can compute and prove the optimal tours for each vehicle individually. If the assignment problem is included as well, we seem to get the optimal solutions but we are not able to prove the optimality within several hours and even several days using Xpress-MP [41]. The basic reason is that the lower bound of this minimization problem improves only very slowly or does not change at all, even when the internal cuts of Xpress-MP (Release 13.26) are used.

4.3.1.7 Tightening the Model Formulation

Here we discuss some valid inequalities (cuts) which can help to decrease the number of arcs. At first, the model formulation can be essentially tightened and the solution time can be significantly reduced by eliminating impossible or unlikely arcs from the set $\mathcal{A}$. For deterministic time window constraints which cannot be relaxed it can be worked out exactly which arcs can be eliminated. As in the problem we are interested in the temporal constraints can be relaxed or slightly violated we can at best argue with some probabilities.

For instance, if orders arrive in 5 minute sequences, it is not likely that a vehicle will ever drive from a node n_1 with target time 8:50am to another node n_2 with target time 3:30pm. Thus, the variable $\delta_{n_1 n_2 v}$ needs not to be generated or can be set to zero. This observation leads to the introduction of a forward cut parameter, Δ^F. Similarly, it is not likely that an arc exists from a node with target time 3:30pm to another node with target time 9:50am. Therefore, we introduce a backward cut parameter, Δ^B. The parameters Δ^F and Δ^B should be chosen with great care. Otherwise the problem can easily become infeasible.

Furthermore, it is obvious that the node connection (j_i, i) needs not to be considered because a vehicle will not drive from a delivery point back to the corresponding pickup node. Let us summarize all rules for eliminating arcs. Arcs corresponding to the variables $\delta_{n_1 n_2 v}$ need not to be generated or can be eliminated by setting $\delta_{n_1 n_2 v} = 0$, if

- the arc represents driving back from a delivery to a pickup point, *i.e.*, n_2 is a pickup node and n_1 is the corresponding delivery node;

- $T_{n_2}^T \geq T_{n_1}^T + \Delta^B$, *i.e.*, if the target time, $T_{n_1}^T$, of node n_1 is much smaller (by value Δ^B) than the target time, $T_{n_2}^T$, of node n_2;

- $T_{n_2}^E + \Delta^F \leq T_{n_1}^E$, *i.e.*, pickup time of node n_2 is much smaller (by value Δ^F) than the pickup time of node n_1;

- n_1 is a destination depot;

- n_1 is a source depot and n_2 is a delivery node;

- n_1 is a pick-up node and n_2 is a destination depot;

- n_2 is a source depot.

Furthermore, special cuts can be applied to tighten the model. Especially, for cars which have only space for two lying passengers, the cuts

$$\delta_{niv} \le C_v^L - p_{nv}^L, \quad \forall n \in \mathcal{N}, \quad \forall i \in \mathcal{P}, \quad \forall v \in \mathcal{V} \qquad (4.3.40)$$

turn out to be very useful and represent the fact that a vehicle should not drive to another pickup point if the capacity is already completely used.

The same holds for orders with patients in wheelchairs, where $C_v^W \le 2$:

$$\delta_{niv} \le C_v^W - p_{nv}^W, \quad \forall n \in \mathcal{N}, \quad \forall i \in \mathcal{P}, \quad \forall v \in \mathcal{V} \; . \qquad (4.3.41)$$

For single-vehicle problems the model can be tightened if a feasible tour is known from applying the heuristic approaches described in Section 4.4.2. In that case, the total lateness, s_L, can be introduced as a new variable and is bounded by

$$s_L := \sum_{v \in \mathcal{V}} \sum_{i \in \mathcal{P}} r_{iv}^{A_2} + \sum_{v \in \mathcal{V}} \sum_{j \in \mathcal{D}} r_{jv}^B \le S_L^+ \quad , \qquad (4.3.42)$$

where S_L^+ is an upper bound obtained, for instance, by a heuristic approach. Of course, one could also restrict the individual sums for pickup and delivery nodes. As is illustrated in the two examples in Section 5.2.3, the effect can be very significant.

4.3.1.8   Concluding Remarks on the Model

The MILP model in its current form described above can solve the problem in favorable cases. However, this is not sufficient for being embedded in a real world application. Therefore, various very special features of the real world problem have not been implemented. This includes carriage, and infectious persons.

4.3.2   A Branch-and-Bound Approach for Solving the Intra-Tour Problem

For certain objective functions, a recursive approach in the branch-and-bound style can be used. This applies to objective functions which can be built-up cumulative when constructing a tour from a given set of nodes by adding nodes to it in sequential manner. For computing the objective function value when the tour has reached a size of n nodes, it is sufficient to know only the objective function values of the previous nodes and adding some extra amount for the current node, n, but nothing from the subsequent levels. The objective functions we consider in this book fulfill this condition, for instance, objective functions like $z = c^D + c^P + c^L + c^N$ (see page 72).

Let us consider the intra-tour optimization problem for one vehicle and n nodes $\mathcal{N}_n := \{n_1, n_2, \ldots, n_n\}$. The node n_s represents the source depot of the vehicle; the index representing the vehicle is dropped. The optimization problem is formally broken into a recursive scheme[3] of minimization problems depending on a smaller number of nodes. The overall objective is to determine the optimal sequence of the node set $\{n_1, n_2, \ldots, n_n\}$. In offline planning, the vehicle leaves the source depot, n_s, at time t_0 (usually leading to zero delay for the first pickup node). In an online case, t_0 may represent the current time T_*. The initial value of the objective function value is $z_0 = 0$. Finally, we need to track the load of the vehicle; the vector $\mathbf{p}$ (initially set to zero) serves this purpose. After leaving n_s, the vehicle can drive to any of the nodes $\mathcal{N}_n$. We express this by $n_s(n_1, n_2, \ldots, n_n)$, to be read as "leave from n_s, and the sequence of the node in "()" is free". The top-node eventually will have the value

$$z_n = f_n(t_0, z_0, \mathbf{p}_0, n_s; \mathcal{N}_n) = \min_{k=1,\ldots,n} \{f_{n-1}(t_0, z_0, \mathbf{p}_0, n_k; \mathcal{N}_{n-1,k})\} \quad ,$$

with

$$\mathcal{N}_{n-1,k} := \mathcal{N}_n \backslash \{n_k\} \quad ,$$

and $f_{n-1}(t_0, z_0, \mathbf{p}_0, n_k; \mathcal{N}_{n-1,k})$ is the objective function value in case a vehicle drives from the top-node to node k.

The functions $f_m(t_{n-m}, z_{n-m}, \mathbf{p}_{n-m}, n_k; \mathcal{N}_{mk})$ are computed recursively for $m = n - 1, \ldots, 2$ as

$$f_m(t_{n-m}, z_{n-m}, \mathbf{p}_{n-m}, n_k; \mathcal{N}_{mk})$$
$$= \min_{k=1,\ldots,m-1} \{f_{m-1}(t_{n-m+1}, z_{n-m+1}, \mathbf{p}_{n-m+1}, n_k; \mathcal{N}_{m-1,k})\}$$

with

$$\mathcal{N}_{m-1,k} := \mathcal{N}_{m,k} \backslash \{n_k\} \quad .$$

The functions f_m, $m = n, \ldots, 2$ contain the following arguments: the time, t_{n-m}, at which the vehicle is ready to leave the source node n_k, the

---

[3]This recursive structure allows to develop a dynamic programming approach as well.

accumulated objective function value, z_{n-m}, associated with the previous $n - m$ nodes, the load vector, $\mathbf{p}$, the source node from which the tour should be continued, and the set of nodes, $\mathcal{N}$, to be put in sequence. The initial value for z_0 is always zero, for t_0 is the current time or a starting time leading to zero delay for the first pickup node. The initial load vector is zero, or if pre-assigned orders are present, to the number of passengers in the car. The tour always finishes by connecting the last node (a delivery node!) to the destination depot of the vehicle. The function, f_m, of level m evaluates the new current time, t_{n-m+1}, and the new objective function value, z_{n-m+1}, when node is added to the tour, and passes those values to the next level function, f_{m-1}.

Now we illustrate this method with the following example.

EXAMPLE 4.3 *Let* $N^O = 2$, *i.e.*, $n = 4$ *nodes need to be sequenced. The nodes 1 and 2 represent the orders, or their pickup nodes, and 3 and 4 the corresponding delivery nodes. Thus, we have* $\mathcal{N}_4 := \{n_1, n_2, n_3, n_4\}$ *and*

$$
\begin{aligned}
& & z_4 &= f_4(t_0, z_0, \mathbf{p}_0, n_s; \mathcal{N}_4) = \min\{ \\
f_3(t_0, z_0, \mathbf{p}_0, n_1; \mathcal{N}_{3,1}^0) &, & \mathcal{N}_{3,1}^0 &= \mathcal{N}_4 \backslash \{n_1\} = \{n_2, n_3, n_4\} \\
f_3(t_0, z_0, \mathbf{p}_0, n_2; \mathcal{N}_{3,2}^0) &, & \mathcal{N}_{3,2}^0 &= \mathcal{N}_4 \backslash \{n_2\} = \{n_1, n_3, n_4\} \\
f_3(t_0, z_0, \mathbf{p}_0, n_3; \mathcal{N}_{3,3}^0) &, & \mathcal{N}_{3,3}^0 &= \mathcal{N}_4 \backslash \{n_3\} = \{n_1, n_2, n_4\} \\
f_3(t_0, z_0, \mathbf{p}_0, n_4; \mathcal{N}_{3,4}^0) &, & \mathcal{N}_{3,4}^0 &= \mathcal{N}_4 \backslash \{n_4\} = \{n_1, n_2, n_3\} \\
& & &\} \; .
\end{aligned}
$$

The functions $f_3(z_0, n_3; \mathcal{N}_{3,3}^0)$ *and* $f_3(z_0, n_4; \mathcal{N}_{3,4}^0)$ *do not need to be evaluated since a tour cannot start with a delivery node. Therefore, the next levels to evaluate are*

$$
\begin{aligned}
& f_3(t_0, z_0, \mathbf{p}_0, n_1; \mathcal{N}_{3,1}^0) &=& \min\{ \\
f_2(t_1, z_1(n_1), \mathbf{p}_1, n_2; \mathcal{N}_{2,2}^1) &, & \mathcal{N}_{2,2}^1 &= \mathcal{N}_{3,1}^0 \backslash \{n_2\} = \{n_3, n_4\} \\
f_2(t_1, z_1(n_1), \mathbf{p}_1, n_3; \mathcal{N}_{2,3}^1) &, & \mathcal{N}_{2,3}^1 &= \mathcal{N}_{3,1}^0 \backslash \{n_3\} = \{n_2, n_4\} \\
f_2(t_1, z_1(n_1), \mathbf{p}_1, n_4; \mathcal{N}_{2,4}^1) &, & \mathcal{N}_{2,4}^1 &= \mathcal{N}_{3,1}^0 \backslash \{n_4\} = \{n_2, n_3\} \\
& & &\} \; .
\end{aligned}
$$

and

$$f_3(t_0, z_0, \mathbf{p}_0, n_2; \mathcal{N}_{3,2}^0) = \min\{$$

$$f_2(t_1, z_1(n_2), \mathbf{p}_1, n_1; \mathcal{N}_{2,1}^2) \quad , \quad \mathcal{N}_{2,1}^2 = \mathcal{N}_{3,2}^0 \backslash \{n_1\} = \{n_3, n_4\}$$

$$f_2(t_1, z_1(n_2), \mathbf{p}_1, n_3; \mathcal{N}_{2,3}^2) \quad , \quad \mathcal{N}_{2,3}^2 = \mathcal{N}_{3,2}^0 \backslash \{n_3\} = \{n_1, n_4\}$$

$$f_2(t_1, z_1(n_2), \mathbf{p}_1, n_4; \mathcal{N}_{2,4}^2) \quad , \quad \mathcal{N}_{2,4}^2 = \mathcal{N}_{3,2}^0 \backslash \{n_4\} = \{n_1, n_3\}$$

$$\} \; .$$

Notice, $\mathcal{N}_{2,4}^1$ *and* $\mathcal{N}_{2,3}^2$ *are not considered either. Level three becomes already the last one to investigate. For example, below the branch derived from* $f_2(t_1, z_1(n_1), \mathbf{p}_1, n_2; \mathcal{N}_{2,2}^1)$ *is displayed:*

$$f_2(t_1, z_1(n_1), n_2; \mathcal{N}_{2,2}^1) = \min\{$$

$$f_1(t_2, z_2(n_1, n_2), n_3; \mathcal{N}_{1,3}^3) \quad , \quad \mathcal{N}_{1,3}^3 = \mathcal{N}_{2,2}^1 \backslash \{n_3\} = \{n_4\}$$

$$f_1(t_2, z_2(n_1, n_2), n_4; \mathcal{N}_{1,4}^3) \quad , \quad \mathcal{N}_{1,4}^3 = \mathcal{N}_{2,2}^1 \backslash \{n_4\} = \{n_3\}$$

$$\} \; .$$

Graphically, the tree and the way we move through it (indicated by the subscripts) can be depicted as shown in Fig. 4.3.4. The subscript indi-

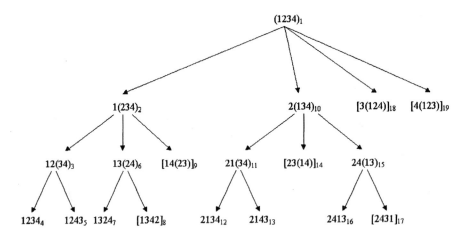

Figure 4.3.4. The branch-and-bound tree for the intra-tour optimization problem

cates the sequence in which the nodes are created. Those nodes in the tree displayed as "[ ]" indicate infeasible nodes; the infeasibility is caused by the delivery node preceding the pickup node. In addition to this, nodes might become infeasible and are pruned because they violate the capacity constraints. The number of tours to be tested derived from (4.3.39) is six; this is identical to the number of feasible tours (fulfilling the precedence constraints) displayed at the bottom line.

Note that z_n can only be computed until the deepest level of the tree is reached on all branches, but never before. As indicated by the subscript in the figure above, we move through the tree in a depth-first type way. While reaching deeper levels of the tree (the top node has level 1) tours are built up by adding more and more nodes leading to increasing times, t, and possibly objective function values, z. If the deepest level of the tree is reached, and the associated node turns out to be feasible, *i.e.*, a feasible tour has been constructed and its associated objective function value, z, has been found and evaluated, the cutoff value, Z_C, is tested for updating. If $z < Z_C$, then $Z_C = z$ (the initial value $Z_C := +\infty$).

Except for eliminations of a priori identified infeasible tours, the tree, in the worst case, represents a full enumeration scheme. Nevertheless, this approach is more efficient, *i.e.*, computationally cheaper than the full enumeration, because nodes and thus whole parts of the tree are pruned when a node turns out to be

1 infeasible, or

2 value-dominated by a previous solution found, or by a pre-set value, Z_C, computed, for instance, as described in Section 4.3.3.6 or,

3 the optimal node identified by the condition $z < 10^{-6}$.

However, before a new node of the tree is evaluated, the algorithm checks a priori the two following obvious conditions:

1 precedence constraints (the pickup node of an order needs to be visited earlier than the corresponding delivery node), and

2 capacity constraints (subtours driving to too many pickup nodes can be avoided).

Additional computing time savings are obtained because the objective function and other quantities associated with a tour are computed successively when nodes are added. A tour with n nodes requires n successive computations of adding a node and updating the accumulated objective function and other quantities. Thus, for two tours, whose last and second last node are just exchanged as in 1234 ánd 1243, the total computational effort for n nodes is $n - 2 + 2 \times 2 = n + 2$. If both tours would have been computed independently, the effort would be $2n$.

This approach is limited mostly by memory requirements. The minimal amount of information we have to store for each node is as follows:

1 node number, n_n

2 parent node, n_p

3 node level, n_l

4 number of nodes in the tour already fixed, n_f

5 position to be filled when the next child node is created, n_r

6 status of the node, n_s

7 the sequence of nodes in the tour, $n_1, n_2, \ldots, n$

8 accumulated time, t

9 accumulated objective function value, z

10 patients load (lying, sitting, wheelchair), p_l, p_s, p_w

To save memory, our program has the option not to store those nodes of the tree which have been pruned or fully evaluated. In that case, the number of nodes of tree's level to be stored is the number, N_L. For our depth-first search, each child-node is stored into the node position corresponding to the level of the tree. If there are no pre-assigned orders, N_L is just the number, n, of nodes. If there are m_0 pre-assigned orders, then $N_L = n - m_0$. Nodes on the deepest level of the tree always show up as a pair, *e.g.*, 142356 and 142365 with 1423(56) as their parent node.

If one needs the tree only for inspection purposes, it suffices to store only the first n_f nodes instead of storing the full tour-node list for each tree-node. This would lead to a memory reduction of a factor two. If it is necessary to store the whole tree, the memory issue becomes problematic. Table 4.1 summarizes some numerical experiments and lists the number, N_T, of nodes in the tree, the number, N_F, of feasible tours evaluated, as well as the CPU time (in seconds) needed to evaluate them.

Table 4.1: Numerical experiments for the branch-and-bound approach

N^O	n	N_T	N_F	*sec*
2	4	19	6	1
3	6	105	13	2
5	10	39,058	2,591	4
6	12	890,272	7,520	8
7	14	2,525,686	16,193	30
8	16	47,706,540	124,563	300

This table illustrates the scaling properties of the approach. However, N_T and N_F depend strongly on whether nodes can be pruned at an early stage.

The algorithm can be summarized as follows.

Algorithm BBA

1 Initialize the top-node;

2 If the node list is not empty, select a new node; otherwise the last created tour is optimal;

3 Create a child node and evaluate the partial tour for this node; then do one of the following substeps

   (a) If that child node is on the deepest level of the tree, evaluate it and thus obtain a new evaluated tour. Evaluate also its next brother to the right; this gives the evaluation of a new tour - all nodes of this level and the parent node can be pruned. Goto Step 2.

   (b) If it is not on the deepest level of the tree then goto Step 3.

Nodes are selected deterministically following a priori depth first rules. If one allows less deterministic schemes, it becomes possible to select the most promising nodes and to continue on those parts of the tree. This strategy is accompanied by the hope to find good solutions providing cutoff bounds.

Let us finally summarize and comment on the basic assumption of this branch-and-bound scheme. The basic assumption is that both the objective function and the constraints can be evaluated sequentially when the sequence of objects is constructed, *i.e.*, it is not necessary to know the complete sequence at the beginning.

The sequencing-scheduling problem involves a set of n distinguishable objects i (orders, jobs, nodes) which need to be sequenced and scheduled, and which are subject to a set of m constraints C_j. Let π be a permutation or sequence of those objects, and let $\tau_k(\pi)$, $k = 1, \ldots, n$, be the set of the first k objects of that sequence π.

We call an objective function, $f(\pi)$, *cumulative* if it can be built up in sequence, *i.e.*, it depends only on the current and previous objects of a given (not yet complete) permutation or sequence:

$$
\begin{aligned}
s_1 &= s_1(\pi) := F_1(\tau_1(\pi)) \\
s_n &= s_n(\pi) := s_{n-1} + F_n(s_{n-1}, \tau_n(\pi)) \\
f(\pi) &= s_n \quad .
\end{aligned}
$$

The function $F_n(s_{n-1}, \tau_n(\pi))$ quantifies the contribution of the first n objects of the sequence to the objective function. A special case is that $F_n(s_{n-1}, \tau_n(\pi))$ depends only on the nth object. The total lateness is a typical example. The current node can be late because there are already previous lateness. The same "cumulative" concept holds for constraints. In our branch-and-bound approach, constraints only need to be evaluated and checked; thus there is no restriction regarding its algebraic character (linear or nonlinear). If a hard constraint (*e.g.*, capacity constraint) is violated, this is communicated back to the branch-and-bound scheme, and the node of the tree is pruned. If a soft constraint (*e.g.*, temporal constraint) is violated, a penalty function is applied and its value is added to the objective function.

Thus, our branch-and-bound method is quite general and is suitable to solve any kind of sequencing-scheduling problem involving cumulative objective functions and cumulative constraints.

4.3.3   Column Enumeration

In this section, for convenience, we use n instead of N^V and r instead of N^O.

4.3.3.1   Motivating a Column Enumeration Approach

On page 82 it had been shown that there exist n^r combinations to assign r orders or requests to n vehicles. We later analyze an example with $n = 8$ vehicles and $r = 6$ orders (in the example, the total number of orders is 14 and there are already eight of them already pre-assigned), *i.e.*, 262,144 combinations. Although, it might be possible to compute all those combinations and select the best one, let us consider another approach which we call *column enumeration approach* (CEA). This approach solves the optimization problem by exploiting its structure. The assignment decisions are decoupled from the routing and scheduling decisions. The idea of the CEA is illustrated by discussing the example below. This example is later described in Chapter 5 in greater depth.

If we need to assign $r = 6$ orders, then we obviously can assign between 0 and 6 orders to each of the n vehicles. There are

$$C_{rj} = \binom{r}{j} = \frac{r!}{(r-j)!j!} \qquad (4.3.43)$$

combinations, if j indicates the number of assigned orders. Thus, as shown further down in Theorem 4.4, the total number of scenarios to evaluate is

$$C_r = n \sum_{j=0}^{r} C_{rj} = n \sum_{j=0}^{r} \binom{r}{j} = 2^r n \quad , \qquad (4.3.44)$$

i.e., for $n = 8$ and $r = 6$, this gives $C_6 = 512$ subsets. Let us assume, that for each subset of orders or *column*, indexed by c, we have solved the individual routing and scheduling problem to optimality, *i.e.*, we know the objective function value Z_c. The optimal tours might have been computed by the branch-and-bound approach described in Section 4.3.2. Let C_6 be the set of all possible assignments, and C_6^* be the set of feasible assignments leading to the objective function value Z_c. Note that $|C_6^*| \leq C_6$. If the vehicles have the same capacity and time availability, symmetry could be reduced by excluding those columns c, and the time to compute Z_c could be saved.

Now we can formulate a set partitioning problem ensuring that all orders are assigned exactly once. In order to do so, we introduce the binary variables, γ_c:

$$\gamma_c = \begin{cases} 1, & \text{if tour } c \text{ is used in an optimal solution} \\ 0, & \text{otherwise.} \end{cases}$$

To illustrate the idea, consider the subsets $\mathcal{O}_1 = \{1, 2, 3\}$, $\mathcal{O}_3 = \{6\}$, and $\mathcal{O}_5 = \{4, 5\}$ of orders which have been identified by solving the set partitioning problem formulated below. For each of them, the optimal tour has been computed given the objective function values Z_1, Z_3, and Z_5. The overall objective function value is just the sum $Z_1 + Z_3 + Z_5$. Other subsets, *e.g.*, $\mathcal{O}_2$ and $\mathcal{O}_4$ have not been selected.

The objective function of the set partitioning problem is thus

$$\min \sum_{c=1}^{C_r} Z_c \gamma_c \quad . \tag{4.3.45}$$

The set partitioning model is simple because only one type of constraint is necessary, namely that one ensuring that each order is covered exactly once. If I_{ci} is an indicator function specifying whether column c contains order i, then the constraints

$$\sum_{c=1}^{C_r} I_{ci} \gamma_c = 1 \quad , \quad \forall i = 1, ..., r \quad , \tag{4.3.46}$$

ensure that each order is contained exactly once. In addition the integrality constraints hold

$$\gamma_c \in \{0, 1\} \quad , \quad \forall c = 1, ..., C_r \quad . \tag{4.3.47}$$

The optimization problem defined by (4.3.45) to (4.3.47) is easy to solve; it finds the optimal solution within seconds. If the problem is solved, the variables γ_c give, via a look-up table, the orders and the vehicle these orders are assigned to. For the larger problems, it might be a problem to store all these data.

The optimal solution derived from 504 columns (note that this number is less than the expected one, 512, because for each of the $n = 8$ vehicles there is one empty subset we are not interested in) columns takes 30 seconds, the Master problem (4.3.45) to (4.3.47) solves in 10 seconds to optimality using Xpress-MP.

With this approach it is also possible to solve the problem to optimality with the same set C of columns if the same set of orders need to be assigned to a smaller set of vehicles. This eliminates the necessity to re-compute the subsets.

4.3.3.2　Comments on Column Generation Techniques

The term *column* usually refers to variables in linear programming parlance. In the context of column generation techniques it has wider meaning and stands for any kind of objects involved in an optimization problem. In vehicle routing problems a column might, for instance, as in the previous subsection represent a subset of orders assigned to a vehicle. In network flow problems a column might represent a feasible path through the network. Finally, in cutting stock problems ([33],[34]) a column represents a pattern to be cut.

Column generation is based on a decomposition into a master problem and a subproblem. In simple cases, such as the ones described by Schrage [76] in the LINDO manual, it is possible to generate all columns explicitly, even within a modeling language. Often, the decomposition has a natural interpretation. If not all columns can be generated, the columns are added dynamically to the problem. Barnhart *et al.* [9] give a good overview on such techniques. A more recent review focussing on selected topics of column generation is [27]. In the context of vehicle routing problems, feasible tours have been added columns as needed by solving shortest path problem with time windows and capacity constraints using dynamic programming [26].

More generally, column generation techniques are used to solve well structured MILP problems involving a huge number, say several hundred thousand or millions, of variables, *i.e.*, columns. Such problems lead to large LP problems, if the integrality constraints of the integer variables are relaxed. If the LP problem contains so many variables (columns) that it cannot be solved with a direct LP solver (revised simplex, interior point method) one starts solving this so-called *master problem* with a small subset of variables yielding the *restricted master problem*. After the restricted master problem has been solved, a pricing problem is solved to identify new variables. This step corresponds to the identification of a non-basic variable to be taken into the basis of the simplex algorithm and coined the term *column generation*. The restricted master problem is solved with the new number of variables. The method terminates when the pricing problems cannot identify any new variables. The most simple version of column generation is found in the Dantzig-Wolfe decomposition [24].

Gilmore and Gomory ([33],[34]) were the first who generalized the idea of dynamic column generation to an integer programming (IP) problem: the cutting stock problem. In this case, the pricing-problem, *i.e.*, the subproblem, is an IP problem itself - and one refers to this as a *column generation algorithm*. This problem is special as the columns generated when solving the relaxed master problem are sufficient to get the optimal integer feasible solution of the overall problem. In general this is not so. If not only the subproblem, but also the master problem involves integer variables, the column generation part is embedded into a branch-and-bound method: this is called a *branch-and-price algorithm*. Note that during the branching process new columns are generated; therefore the name *branch-and-price*.

The reader might ask why we do not use this superior techniques but use the rather special column enumeration approach. We have done so for two reasons:

- in the online scenarios we are interested, we need only to solve small problems instances, where column enumeration is sufficient enough and

- column enumeration is much easier to implement.

4.3.3.3   Developing a Column Enumeration Approach

If it is possible to generate all columns explicitly rather than implicitly by a set of linear inequalities, a column enumeration approach [9] might be possible. The limits of this approach when applied to our problem are discussed later. Now we describe how to generate the complete set, $\mathcal{C}_r$, of all columns, *i.e.*, subsets of orders $i \in \mathcal{O}$, $r = |\mathcal{O}|$, assigned to a set of vehicles, $v \in \mathcal{V}$. Let $\mathcal{C}_r$ be the union of the sets, $\mathcal{C}_{rv}$, *i.e.*, $\mathcal{C}_r = \cup_{v=1...n}\mathcal{C}_{rv}$, where $\mathcal{C}_{rv}$ contains the subsets of orders assigned to vehicle v. Note that $\mathcal{C}_{rv}$ contains all subsets containing 1, 2, or r orders assigned to vehicle v. Above it was claimed that $C_r = 2^r n$.

THEOREM 4.4 *Let r be the number of orders and n the number of vehicles. The cardinality $C_r := |\mathcal{C}_r|$, of the set $\mathcal{C}_r$, of all generated columns* *is*

$$C_r = n \sum_{j=0}^{r} C_{rj} = n \sum_{j=0}^{r} \binom{r}{j} = 2^r n \quad .$$

Proof. The proof is based on some elementary combinatorial properties. The factor n is obvious. It just reflects the fact that the same combinations of orders can be assigned to each vehicle. The coefficient C_{rj} represents the number of combinations of how to extract j elements from a set of r distinct elements, and can be found in any book on combinatorics. The sum $\sum_{j=0}^{r} C_{rj}$ represents the number of ways in which a population of r elements can be partitioned into subsets if the size of the subsets is allowed to be any number $j = 0, 1, \ldots, r$. Noting that

$$(x+1)^n = \sum_{j=0}^{r} \binom{r}{j} x^{r-j} \quad , \tag{4.3.48}$$

it follows that, if we set $x = 1$ in (4.3.48), we obtain with $1^{r-j} = 1$

$$\sum_{j=0}^{r} \binom{r}{j} = 2^r \quad (q.e.d).$$

∎

We have implemented two different approaches to generate the set $\mathcal{C}_r$. The first approach works as follows. As there exist 2^r subsets for each vehicle, the construction of the sets explores the binary representation of the number $k = 0, 1, \ldots, 2^r - 1$. Note that each number $n \in \mathbb{N}$ has a unique binary representation. Let the orders i be indexed as $1, \ldots, r$. Since $k < 2^r$, the binary representation of k is given by

$$k = \sum_{\nu=0}^{r} A_{k\nu} 2^{\nu} \quad , \quad A_{k\nu} \in \{0, 1\} \quad ,$$

where $A_{k\nu}$ is the coefficient representing the contribution of 2^{ν} to k.

Assuming for the moment that all vehicles v have the same properties w.r.t. time availability, we define the subsets $\mathcal{C}_{rv}^{k}$ as

$$\mathcal{C}_{rv}^{k} := \{b + 1 : 1 \leq b \leq r - 1 \wedge A_{kb} = 1\} \quad , \quad \forall v \in \mathcal{V} \quad ,$$

which means that the binary coefficients of k select the orders uniquely. Let us illustrate this by two examples

$$\begin{aligned} k &= 10 = 2^1 + 2^3 \rightarrow \mathcal{C}_{6v}^{10} = \{2, 4\} \\ k &= 43 = 2^0 + 2^1 + 2^3 + 2^5 \rightarrow \mathcal{C}_{6v}^{43} = \{1, 2, 4, 6\} \quad . \end{aligned}$$

Those numbers of the orders generated in such a way are subject to a permutation mapping assigning the real names of the orders. If the list of orders is {A1,B5,A4,C2}, then, for $k = 10$, the result is C_{6v}^{10}={B5,C2}. If the vehicles differ in their time availability, some sets C_{rv} can be eliminated if they contain orders which drastically violate the availability. For example, if a vehicle is only available in the morning, subsets containing orders with requested pickup times at 4pm hours are completely infeasible. Note that one would not eliminate them, if they violate a time availability only slightly. To make this point clearer, consider an example in which the vehicle is available from 8am to 12am, but the requested pickup time is 7:55am. Strictly speaking, this order is not compatible with the availability of that vehicle. But if one accepts a small delay of only five minutes, then this order can be assigned to this vehicle. Therefore, the elimination of columns applies more to offline optimization scenarios covering several hours, while online scenarios have a much smaller time range.

The number, $N(C_r)$, of operations to generate all columns, C_r, is related to the computation of the binary representation of the numbers $k = 1, \ldots, 2^r - 1$ and thus scales with

$$N(C_r) \sim 2^r (r - 1) \quad . \tag{4.3.49}$$

As discussed above in the current section there might be a need to restrict the size of the columns to at most, say, m orders. If $m < r$, we generate all subsets but just discard those having more than m orders. For larger values of r it seems to be recommended to use a different variant to generate the subsets, which considers m right from the beginning; otherwise the effort to generate all subsets might be too large.

The idea of the second approach is to generate the subsets successively for $m' = 1, \ldots, m$. For $m' = 1$ it is clear how to proceed and to initialize the procedure. There are r subsets $\{i\}$ of size $m' = 1$ each containing one order $i = 1, \ldots, r$. For $m' \geq 2$ we loop over all subsets of size $m' - 1$. Let i_{m-1}^* be the largest order in a subset, S_{m-1}^*, selected from level m'. Then we loop over all orders $i > i_{m-1}^*$, adding $\{i\}$ to S_{m-1}^*, *i.e.*, $S_m^* = S_{m-1}^* \cup \{i\}$, and thus generate $r - i_{m-1}^*$ subsets of size m. Below this idea is illustrated by an example for $r = 4$ and $m = 3$:

$s = 1:$	{1} , {2} , {3} , {4}
$s = 2:$	{1,2}, {1,3}, {1,4}
	{2,3}, {2,4}
	{3,4}
$s = 3:$	{1,2,3}, {1,2,4}
	{1,3,4}
	{2,3,4}

As an example, let us derive the level $m' = 3$ subsets from the subset $\{1,2\}$. In this case, we have $i^*_{m-1} = 2$. Thus, looping over $i \geq i^*_{m-1} + 1$ adds $\{3\}$ and $\{4\}$ to the subset $\{1,2\}$, *i.e.*, $\{1,2\}$ generates the $r - i^*_{m-1} = 2$ subsets $\{1,2,3\}$ and $\{1,2,4\}$ and so on (see Fig. 4.3.5).

In order to avoid storing the subsets in memory, it is more efficient to generate them in a depth-first type fashion in the following sequence indicated by the subscript (actually, it is a lexicographical order if subsets of the same level are considered). This algorithm to generate the subsets

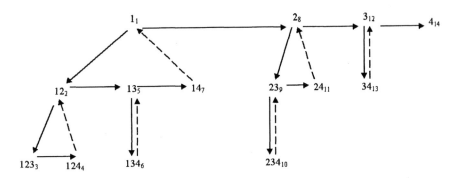

Figure 4.3.5.   Generating all subsets in the column enumeration approach

is, especially for $m < r$, more efficient than the one based on the binary representation.

Another complication arises when generating the columns for vehicles which have already some passengers on board, *i.e.*, m_{0v} out of $m_{0v} + r$ orders have already been assigned to vehicle v. Consider the case that already two orders have been assigned to vehicle v, and that the size of the column should not exceed $m = 4$ orders.

To give an example, we take the set of freely assignable orders to be

$$\mathcal{O} = \{A1, A2, A3, A4, A5, A6\}$$

while the orders A7 and A8 have already been assigned to vehicle v. As $m_{0v} = 2$ and $m = 4$, only two additional orders, *i.e.*, $m - m_{0v}$ orders can be assigned. All columns contain the orders A7 and A8. Then, 0, 1, or 2 orders out of 6 can to be added. Note, this time, we need to add the empty set corresponding to $k = 0$ as well. For this special case there exist $C_{60} + C_{61} + C_{62} = 1 + 6 + 15 = 22$ columns for vehicle v.

In there are no pre-assigned orders for vehicle v, then for $m = 4$ there exist $C_{61} + C_{62} + C_{63} + C_{64} = 6 + 15 + 20 + 15 = 56$ columns (note that this time the empty column set is not considered) for vehicle v instead of 512, *i.e.*, we save almost 90% of all columns!

4.3.3.4 Summarizing the Column Enumeration Approach

Algorithm CEA

1 Explicit generation of all columns C_{rv} as outlined before; followed by a simple feasibility test *w.r.t.* the availability of the vehicles.

2 Solving the routing-scheduling problem for all columns C_{rv} using the branch-and-bound approach described in Section 4.3.2 (the optimal objective function values, Z_c or Z_{cv}, respectively, and the associated routing-scheduling plan are stored).

3 Solving the set partitioning model (4.3.50)-(4.3.53):

$$\min_{\gamma_{cv}} \sum_{c=1}^{C_{rv}} \sum_{v=1}^{N^V} Z_{cv} \gamma_{cv} \quad , \tag{4.3.50}$$

subject to

$$\sum_{c=1}^{C_{rv}} \sum_{v=1}^{N^V} I_{cvi} \gamma_{cv} = 1 \quad , \quad \forall i = 1, ..., r \quad , \tag{4.3.51}$$

ensuring that each order is contained exactly once, the inequality

$$\sum_{c=1}^{C_{rv}} \gamma_{cv} \leq 1 \quad , \quad \forall v \in \mathcal{V} \quad , \tag{4.3.52}$$

ensuring that at most one column can exist for each vehicle, and the integrality conditions

$$\gamma_{cv} \in \{0, 1\} \quad , \quad \forall c = 1, ..., C_{rv} \quad . \tag{4.3.53}$$

Unlike the more generic form of the optimization problem defined by (4.3.45) to (4.3.47), the binary variables γ_{cv} and the indicator function I_{cvi} now depend on the vehicle index v:

$$\gamma_{cv} = \begin{cases} 1, & \text{if column } cv \text{ is selected} \\ 0, & \text{otherwise.} \end{cases} \quad ,$$

and

$$I_{cvi} = \begin{cases} 1, & \text{if column } cv \text{ includes order } i \\ 0, & \text{otherwise.} \end{cases} \quad .$$

This allows us to determine easily to which vehicle an order is assigned. However, note that not all combinations of index pairs $\{c, v\}$ exist; each c corresponds to exactly one v, and vice versa. But much more important, this formulation allows us to find optimal solutions with the defined columns for a smaller number of vehicles. The objective function and the set partitioning constraints are just modified by replacing

$$\sum_{v=1|v\in\mathcal{V}}^{N^V} \longrightarrow \sum_{v=1|v\in\mathcal{V}_*}^{N^V} \quad ,$$

the equations (4.3.51) by

$$\sum_{c=1}^{C_{rv}} \sum_{v=1|v\in\mathcal{V}_*}^{N^V} I_{icv}\gamma_{cv} = 1 \quad , \quad \forall i = 1, ..., r \quad ,$$

and the inequalities (4.3.52) by

$$\sum_{c=1}^{C_{rv}} \gamma_{cv} \leq 1 \quad , \quad \forall v \in \mathcal{V}_* \quad ,$$

where $\mathcal{V}_* \subset \mathcal{V}$ is a subset of the set $\mathcal{V}$ of all vehicles. Alternatively, if it is not pre-specified which vehicles should be used but it is only required that not more than N_*^V vehicles are used, then the inequality

$$\sum_{c=1}^{C_{rv}} \sum_{v=1|v\in\mathcal{V}}^{N^V} \gamma_{cv} \leq N_*^V \tag{4.3.54}$$

is imposed.

4 Re-constructing the complete solution and extracting the complete solution from the stored optimal solutions for the individual columns.

4.3.3.5   Complexity Aspects and Computational Issues

The complexity and thus, the numerical efficiency and also the size of the problems which can be solved in reasonable time on a standard computer (Pentium III, 750 MHz, 512 MB memory, 20 GB harddisc) depends on the following parameters:

1 the number of vehicles, n,

2 the number of orders to be assigned, r, and

3 the maximal number, m, of orders to be assigned to an individual vehicle.

Next we discuss how memory and computational time grow as functions of these parameters.

The memory requirements of our CEA growths linearly in the number, n, of vehicles, as the number of columns generated scales with n. Thus, the computational time to evaluate all the columns is also proportional to n. From our computational experiments for the number of vehicles such that $5 \leq n \leq 20$, the complexity of the set partitioning model does not grow very fast. Using the commercial MILP solver Xpress-MP [41], the set partitioning problem is solved to optimality within seconds.

The number, C_r, of columns to be generated for r orders, is 2^r. Note that r is the number of free orders to be assigned. Pre-assigned orders increase the number of orders contained in the columns but not the number C_r. Typical scenarios to be solved are $r = 10$ leading to $C_r = 1,024$, while $C_{15} = 32,768$ and $C_{20} = 1,048,576$. The computational effort to generate all subsets growths with $2^r(r-1)$ as discussed before near (4.3.49). This requires solving huge set partitioning problems. The largest set partitioning problems we have solved so far included 12,030 columns, *i.e.*, 12,030 binary variables. It seems possible to solve larger set partitioning problems but then more subproblems, and even worse, larger subproblem have to be solved. However, the subproblems become smaller in size when only up to m orders are allowed in each column. Then, only

$$C_r^m = n \sum_{j=0}^{m} C_{rj} \qquad (4.3.55)$$

columns are considered and need to be stored; let us denote the associated set of all columns by $\mathcal{C}_r^m$, *i.e.*, $C_r^m = |\mathcal{C}_r^m|$. The number m may be derived from the requirement that approximately the same number of orders should be assigned to each of the available vehicles, *i.e.*,

$$m \approx \frac{r}{n} + 2 \quad . \qquad (4.3.56)$$

Note that we added 2 to the approximate average value to be safe and not to miss an optimal solution.

If m is significantly smaller than r, the reduction is significant as well, *e.g.*, for $r = 20$ and $m = 5$ we are talking about

$$n \sum_{j=0}^{5} \binom{20}{j} = n(1 + 20 + 190 + 1,140 + 4,845 + 15,504) = 21,700n$$

rather than $2^{20}n \approx 10^{0.3 \cdot 20}n = 10^6 n$. This example might be realistic when assigning 20 orders to 10 vehicles; some might have already some pre-assigned orders.

For each column the optimal objective function value and its optimal tour sequencing need to be stored. Assuming that each tour node just requires 1 Byte to be stored, the total memory requirement (in Bytes) is (the proof follows after Theorem 4.5)

$$M_r = n \sum_{j=0}^{r} (2j) C_{rj} = 2n \sum_{j=0}^{r} j \begin{pmatrix} r \\ j \end{pmatrix} = 2nr2^{r-1} = nr2^r \quad . \quad (4.3.57)$$

With $1\text{kB} = 1024 = 2^{10}$ Bytes we get the memory requirements, M_r^*, in MBytes

$$M_r^* = \frac{M_r}{(2^{10})^2} = nr2^{r-20} \quad .$$

Thus, for $n = 10$ vehicles we get the following results in MBytes

r	13	15	20	25
M_r^*	1.016	4.668	200	8,000

This indicates that, in terms of memory, it is hard to use the CEA beyond $r \approx 20$. However, the size of the partitioning models to be solved limits the approach probably already to $r \approx 15$. As mentioned above, the problem may become significantly smaller, if not all subsets but only those up to a size of $m \cdot$ orders are considered.

The real limiting size is that of the largest routing-scheduling problem we are able to solve exactly. Thus, the critical quantity is the ratio, N_v^O, of orders per vehicle to be sequenced. Thus, if the fleet has 20 vehicles, it is still possible to solve problems with approximate $20N_v^O$ orders.

THEOREM 4.5 *For arbitrary $r \in \mathbb{N}$, $j \in \mathbb{N}_0$ and $j \leq r$ the following equation holds*

$$\sum_{j=0}^{r} j \begin{pmatrix} r \\ j \end{pmatrix} = r2^{r-1} \quad .$$

Proof. *We prove the equation by induction over r.*

$r = 1$.
$\sum_{j=0}^{1} j \begin{pmatrix} 1 \\ j \end{pmatrix} = 0 + 1 = 1 = 1 \cdot 2^0$. Thus, for $r = 1$ the statement is true. Now let us consider the induction step. Suppose the assumption is true for arbitrary r, i.e.,

$$\sum_{j=0}^{r} j \begin{pmatrix} r \\ j \end{pmatrix} = r2^{r-1}. \quad (4.3.58)$$

Let us calculate it for r+1:

$$\sum_{j=0}^{r+1} j \binom{r+1}{j} = \sum_{j=0}^{r} j \binom{r}{j} + \sum_{j=1}^{r+1} j \binom{r}{j-1} =$$

$$r2^{r-1} + \sum_{j=0}^{r}(j+1)\binom{r}{j} = r2^{r-1} + \sum_{j=0}^{r} j \binom{r}{j} + \sum_{j=0}^{r}\binom{r}{j} =$$

$$r2^{r-1} + r2^{r-1} + 2^r = r2^r + 2^r = (r+1)2^r \quad (q.e.d). \quad \blacksquare$$

4.3.3.6 Numerical Improvements

In order to compute exact solutions for the overall problem, the routing-scheduling problem also needs to be solved to optimality. As outlined in Section 4.3.2 the number of orders, or better, nodes, to be sequenced is the critical quantity. Thus, the question is whether it is possible to compute an optimal solution without the need to evaluate all columns with a higher number of orders. The answer is yes and the approach works as follows.

Let $z_{rm} = z(\mathcal{C}_r^m)$ be the optimal objective function value derived from the sets of columns, $\mathcal{C}_r^m$, defined near (4.3.55). Moreover, the cutoff value, Z_C, is initialized to $Z_C = z_{rm}$. The size, m, of the subsets generated, leading, in the presence of pre-assigned orders, to columns with at most $m+1$ orders, should be chosen as small as possible. If we assume that the number, r, of free orders to be assigned is $r \geq 6$, then a reasonable choice is

$$m = \min\{3, r_*\} \quad , \quad r_* := \max\{N_v^* : v = 1, \ldots, n\} \quad ,$$

where r_* is the maximum number of pre-assigned orders in the current vehicle set.

Now we generate all the subsets needed to obtain the column of $\mathcal{C}_r^{m+1}$. Note that $\mathcal{C}_r^m \subset \mathcal{C}_r^{m+1}$ and for each $\mathcal{C} \in \mathcal{C}_r^{m+1}$ there exists a number $k \in \{1, \ldots, r\}$ and a column $\mathcal{C}' \in \mathcal{C}_r^m$ such that

$$\mathcal{C}' = \mathcal{C} \setminus \{k\} \quad ,$$

i.e., for each column, $\mathcal{C}$, of the larger set, $\mathcal{C}_r^{m+1}$, of columns we can find a column, $\mathcal{C}'$, in the previous set, $\mathcal{C}_r^m$, of columns. Let z' denote the optimal objective function value of the routing-scheduling problem obtained for column $\mathcal{C}'$. We do not evaluate those columns, $\mathcal{C} \in \mathcal{C}_r^{m+1}$, for which $z' > Z_C$ because such columns $\mathcal{C}$ can never be part of an optimal solution of the set partitioning model (4.3.50) and (4.3.53). However, we store the column and its associated objective function value to $z_{rm+1} = Z_C + 1$ to be able to use it as a reference column in further iterations. For those new columns remaining to be evaluated, it is also passed to the branch-and-bound approach described in Section 4.3.2.

4.3.3.7 Column Enumeration Coupled to Heuristics

To fully explore the advantages of our CEA, the critical parameters r and N_v^O should be matched to each other. While it is possible to solve the set partition problem with up to $r \approx 15$ or a little more, the branch-and-bound approach becomes inefficient for $N_v^O > 10$. The gap between 10 and 15 can be bridged by solving the routing-scheduling problem not the approach described in Section 4.3.2 but by solving it using the sequencing heuristic described in Section 4.4.2. Of course, this cannot be claimed to be the optimal solution; it is rather a heuristic in itself which we could call IT3. However, SH has been shown to produce good quality solutions for those scenarios for which we have been able to compute the optimal solution via the branch-and-bound approach. Thus, we hope that this heuristic IT3 produces good solutions for these scenarios as well.

4.4. Construction and Improvement Heuristics

Here we present heuristics to solve the problem described in Section 4.3.1. The main purpose of the heuristic is to solve larger instances of the problem. Each heuristic consists of two parts: construction of tours, and improvement of existing tours. The obtained tours are improved later. After assigning orders to vehicles, the next step is to sequence the pickup and delivery nodes associated with these orders in sets $\mathcal{O}_v$ and to generate initial tours. The improvement part has two phases: first, improving the routing and scheduling of each tour separately (the sequence of nodes in each set $\mathcal{O}_v$ is changed), and, second, pickup-delivery node pairs are moved from $\mathcal{O}_{v_1}$ to $\mathcal{O}_{v_2}$.

4.4.1 Construction Heuristics

All construction heuristics we propose consist of several stages:

1 assigning orders to vehicles;

2 distributing the orders not assigned to any vehicle during step 1;

3 sequencing the pickup and delivery nodes of orders and generating a tour.

The first stage is common to all heuristics. The basic idea is to distribute the orders of set $\mathcal{O}$ among N^V vehicles (*i.e.*, assign them to sets $\mathcal{O}_v$) in such a way that $\mathcal{O} = \cup_{v=1}^{N^V} \mathcal{O}_v, \mathcal{O}_i \cap \mathcal{O}_j = \emptyset, i \neq j$. Using a set of N^B time bins with a typical value of $N^B = 24$, it is possible to operate in one-hour time slices (for instance, if the pickup time of an order is 8:55am, then the active time slice is 9; if 9:05am, then the active time slice is 10; if

4:10pm, then the active time slice is 17). First, the orders are sorted in ascending order of pickup times. Order i is assigned to set $\mathcal{O}_v$, if the following conditions hold simultaneously:

- the active time slice k of order i falls into the time window during which vehicle v is available;

- vehicle v has not more than N^O/N^V orders (average per vehicle) in time slice k;

- vehicle v has no orders with similar pickup times (*i.e.*, the difference in pickup times not more than, for instance, 5 minutes);

- the capacity constraints are not violated for vehicle v, *i.e.*, in time slice k values C_v^L, C_v^W or C_v^S are not exceeded.

If there are several vehicles satisfying these conditions, then the vehicle, v_b, that has the lowest number of orders in time slice k, is chosen. Finally, N^V disjunct sets, $\mathcal{O}_v$, of orders are produced. Orders that are not assigned using the heuristic described above, are collected in the set $\mathcal{O}^U$ of unassigned orders. The orders $i \in \mathcal{O}^U$ are treated as additional orders to be assigned during stage 2. As orders involving lying passengers may lead easier to infeasibilities, such orders are distributed earlier then all other types of orders.

At stage 2, orders $o \in \mathcal{O}^U$ have to be assigned to the existing tours $\mathcal{T}_v$. Again, we sort them *w.r.t.* increasing pickup times. Further there are two possibilities to distribute the orders. The simpler heuristic, DU, distributes the orders uniformly to the available vehicles irrespective of all constraints, *i.e.*, the first order from the set $\mathcal{O}^U$ is assigned to the first vehicle, the second order to the second vehicle and so on. In this case, the capacity constraints can be violated; this problem is discussed later. The second heuristic, DB, distributes the remaining orders to the vehicles, which would suffer less, *i.e.*, each order is added at the best place of the tour, which gives minimal sum of delays, S_v^D. DB is usually combined with stage 3.

Finally, stage 3 can be performed using two heuristics, CONH1 and CONH2 described below.

4.4.1.1 Heuristic CONH1

This heuristic arranges the orders in each set $\mathcal{O}_v$ according to increasing pickup times of the orders, *i.e.*, $T_{i_{o_m}}^E \leq T_{i_{o_{m+1}}}^E$ for all $m = 1, \ldots, N_v^O - 1$, where i_{o_m} is the pickup node of order o_m (furthermore, we define j_{o_m} to be its delivery node). If the target time for order o_m

is set for delivery instead of pickup, then the pickup time is calculated by the formula: $T^E_{i_{o_m}} = T^L_{i_{o_m}} - \Delta_T$. The initial tour for each vehicle v is constructed as follows: the vehicle just leaves its source depot, drives to the pickup node of the first order, delivers the patient at its delivery node, drives to the pickup node of the next order, and so on, *i.e.*,:

$$s_v \rightarrow i_{o_1} \rightarrow j_{o_1} \rightarrow i_{o_2} \rightarrow j_{o_2} \rightarrow \ldots \rightarrow i_{o_{N^O_v}} \rightarrow j_{o_{N^O_v}} \rightarrow d_v$$

respectively, the tour is defined as a sequence of nodes

$$\mathcal{T}_v := n_0 \rightarrow n_1 \rightarrow n_2 \rightarrow \ldots \rightarrow n_{2N^O_v - 1} \rightarrow n_{2N^O_v} \rightarrow n_{2N^O_v + 1} \quad .$$

The target times for such tours may be violated but possibly not too much. Obviously, the capacity constraints and the Δ_T limit are satisfied. The quality of the tour with respect to the temporal constraints is qualified by the maximum time deviation, D_v, and the total lateness S^D_v.

If, during the second stage, orders originally not distributed are assigned to vehicle v uniformally, some orders in a tour may have similar pickup times (not differing by more than, say, $\varepsilon = 0.05$, in hours). They are integrated into the tour as follows (let, for instance, the orders o_1, o_2, o_3 have the same pickup times):

$$s_v \rightarrow \ldots \rightarrow i_{o_1} \rightarrow i_{o_2} \rightarrow i_{o_3} \rightarrow j_{o_1} \rightarrow j_{o_2} \rightarrow j_{o_3} \rightarrow \ldots \rightarrow d_v$$

If vehicle v has pre-assigned orders, at first the pre-assigned patients have to be delivered. Let o_{p_1} and o_{p_2} be pre-assigned orders with $T^E_{i_{o_{p_1}}} \leq T^E_{i_{o_{p_2}}}$, then the nodes are sequenced as depicted:

$$s_v \rightarrow j_{o_{p_1}} \rightarrow j_{o_{p_2}} \rightarrow i_{o_1} \rightarrow j_{o_1} \rightarrow i_{o_2} \rightarrow j_{o_2} \rightarrow \ldots \rightarrow i_{o_{N^O_v}} \rightarrow j_{o_{N^O_v}} \rightarrow d_v$$

4.4.1.2   Heuristic CONH2

For this heuristic, all orders in each set, $\mathcal{O}_v$, have to be sorted according to increasing pickup times. Let order $o_1 \in \mathcal{O}_v$ have the earliest pickup time and order $o_{|\mathcal{O}_v|}$ have the latest pickup time. At the beginning, tour $\mathcal{T}_v$ consists of only one order, o_1:

$$s_v \rightarrow i_{o_1} \rightarrow j_{o_1} \rightarrow d_v \qquad (4.4.1)$$

Each subsequent order o_m, starting with o_2, has to be added to $\mathcal{T}_v$. The pickup node i_{o_m} and the delivery node j_{o_m}, associated with order o_m, will be inserted at the best positions in $\mathcal{T}_v$, *i.e.*, to the positions which gives minimal value for S^D_v. This will be done by using full enumeration illustrated in the example below. Suppose order o_2 has to be added. As the delivery node of an order is visited later than the corresponding pickup node, only six potential tours are possible:

$$T_v^1 : s_v \rightarrow i_{o_1} \rightarrow j_{o_1} \rightarrow i_{o_2} \rightarrow j_{o_2} \rightarrow d_v$$

$$T_v^2 : s_v \rightarrow i_{o_1} \rightarrow i_{o_2} \rightarrow j_{o_1} \rightarrow j_{o_2} \rightarrow d_v$$

$$T_v^3 : s_v \rightarrow i_{o_1} \rightarrow i_{o_2} \rightarrow j_{o_2} \rightarrow j_{o_1} \rightarrow d_v$$

$$T_v^4 : s_v \rightarrow i_{o_2} \rightarrow j_{o_2} \rightarrow i_{o_1} \rightarrow j_{o_1} \rightarrow d_v$$

$$T_v^5 : s_v \rightarrow i_{o_2} \rightarrow i_{o_1} \rightarrow j_{o_2} \rightarrow j_{o_1} \rightarrow d_v$$

$$T_v^6 : s_v \rightarrow i_{o_2} \rightarrow i_{o_1} \rightarrow j_{o_1} \rightarrow j_{o_2} \rightarrow d_v$$

Now we choose that tour T_v^* for which

$$S_v^{D*} := \min \left\{ S_v^{D_p}, 1 \leq p \leq 6 \right\} \quad .$$

Notice, that for some of T_v^* the capacity constraints can be violated. Since capacity constraints are hard, the penalty for violating them is much higher than for violating time constraints. Therefore, in the worst case CONH2 produces the same solution as CONH1. There are slight differences in the construction of the initial tour T_v, if vehicle v has pre-assigned orders $o_{p_1}, o_{p_2}, ..., o_{p_b}$. In this case the initial T_v is constructed as follows:

$$s_v \rightarrow i_{o_{p_1}} \rightarrow i_{o_{p_2}} \rightarrow ... \rightarrow i_{o_{p_b}} \rightarrow j_{o_{p_1}} \rightarrow d_v \quad . \tag{4.4.2}$$

The pickup and delivery nodes of the remaining orders can be inserted at positions which follows after the block $i_{o_{p_1}} \rightarrow i_{o_{p_2}} \rightarrow ... \rightarrow i_{o_{p_b}}$, while corresponding patients are already inside the vehicle v. The heuristic inserts the remaining nodes in T_v, starting with delivery nodes $j_{o_{p_2}}, ..., j_{o_{p_b}}$ of pre-assigned orders. For instance, $j_{o_{p_2}}$ has only two possible insertion positions:

$$s_v \rightarrow i_{o_{p_1}} \rightarrow i_{o_{p_2}} \rightarrow ... \rightarrow i_{o_{p_b}} \rightarrow j_{o_{p_2}} \rightarrow j_{o_{p_1}} \rightarrow d_v \quad ,$$

$$s_v \rightarrow i_{o_{p_1}} \rightarrow i_{o_{p_2}} \rightarrow ... \rightarrow i_{o_{p_b}} \rightarrow j_{o_{p_1}} \rightarrow j_{o_{p_2}} \rightarrow d_v \quad .$$

The following theorem estimates the performance of CONH2.

THEOREM 4.6 *Heuristic* CONH2 *constructs the tours in polynomial time.*

Proof. Let us see how many operations are necessary to construct one tour. Consider the case when set $\mathcal{O}_v$ has no pre-assigned orders. Then $|\mathcal{O}_v| - 1$ orders have to be added to the initial tour. This means that we need to insert $2(|\mathcal{O}_v| - 1)$ nodes. To insert the first node from the set of remaining nodes, there are three possible insertion positions (see 4.4.1), for the second one - four positions, for the third one - five positions, and so on, for the $2(|\mathcal{O}_v| - 1)th$ node there exist $2|\mathcal{O}_v|$ possible insertion positions. In total, it takes

$$3+4+...+2|\mathcal{O}_v| = -3 + \sum_{n=1}^{2|\mathcal{O}_v|} n = \frac{2|\mathcal{O}_v|(2|\mathcal{O}_v|+1)}{2} - 3 = 2(|\mathcal{O}_v|)^2 + |\mathcal{O}_v| - 3$$

operations to construct one tour. This procedure is applied to each vehicle, *i.e.*, in total, N^V times. Notice, that $\max_{v=1,...,N^V}(|\mathcal{O}_v|) = N^O$. Thus, it takes $O(N^O, N^V) \approx N^V (N^O)^2$ operations to construct all tours. Obviously, $O(N^O, N^V)$ is a polynom. If $\mathcal{O}_v$ has pre-assigned orders, the initial tour $\mathcal{T}_v$ has at least two nodes, therefore at most $2(|\mathcal{O}_v| - 1)$ nodes need to be inserted. Thus it takes even less number of operations to construct tours (q.e.d). ∎

After the construction phase of the heuristic, each vehicle has a sequence of nodes, $\mathcal{T}_v$, to be visited. Combining CONH1 or CONH2 with DU or DB there are four different ways to construct tours $\mathcal{T}_v$. The following examples illustrate the construction heuristics.

Table 4.2: Input information about orders

order	T^E	type
o_1	13.50	sitting
o_2	13.75	sitting
o_3	13.75	sitting
o_4	13.76	lying
o_5	14.00	lying

Table 4.3: Input information about vehicles

v	T_v^{VE}	T_v^{VL}
1	13.75	14.20
2	13.75	24.00

EXAMPLE 4.7 *Consider the set of orders presented in Table 4.2 with information about the availability of the vehicle from Table 4.3. Here $N^O = 5$, $N^V = 2$. Let the current time be 13.75 and $C_v^L = 2$, $C_v^S = 5$. Notice that some orders have pickup times earlier than the current time (it means these orders will be certainly delayed).*
After distributing all orders, the sets $\mathcal{O}_v$ looks as follows: $\mathcal{O}_1 = \{o_4, o_5\}$, $\mathcal{O}_2 = \{\emptyset\}$. Orders originally not assigned are collected in the set $\mathcal{O}^U = \{o_1, o_2, o_3\}$. Orders from $\mathcal{O}^U$ will be distributed using DU and DB methods.

EXAMPLE 4.8 *Using the sets $\mathcal{O}_v$ from the previous example, the heuristics* CONH1 *and* CONH2 *lead to the following results for T_v, D_v and S_v^D (in hours).*

DU and CONH1:

v	D_v	S_v^D	T_v
1	0.61	1.78	$i_{o_1} \to j_{o_1} \to i_{o_3} \to j_{o_3} \to i_{o_4} \to j_{o_4} \to i_{o_5} \to j_{o_5}$
2	0.00	0.00	$i_{o_2} \to j_{o_2}$

DU and CONH2:

v	D_v	S_v^D	T_v
1	0.61	1.66	$i_{o_1} \to j_{o_1} \to i_{o_3} \to i_{o_4} \to j_{o_3} \to j_{o_4} \to i_{o_5} \to j_{o_5}$
2	0	0	$i_{o_2} \to j_{o_2}$

DB and CONH1, CONH2:

v	D_v	S_v^D	T_v
1	0.33	1.00	$i_{o_1} \to i_{o_4} \to i_{o_3} \to j_{o_4} \to i_{o_5} \to j_{o_3} \to j_{o_1} \to j_{o_5}$
2	0.00	0.00	$i_{o_2} \to j_{o_2}$

In this case DU *and* DB *distribute orders contained in $\mathcal{O}^U$ in the same way:*
$$\mathcal{O}_1 := \mathcal{O}_1 \cup \{o_1, o_3\},$$
$$\mathcal{O}_2 := \mathcal{O}_2 \cup \{o_2\}.$$

As expected, the method CONH2 produced a better solution. The quality of the constructed tours with $N_v^O \leq 2$ is estimated by the following proposition.

PROPOSITION 4.9 *Let T_v be an arbitrary tour. If T_v consist of two orders o_1 and o_2, then a solution produced by* CONH2 *is optimal.*

Notice that this proposition cannot be extended to $N_v^O \geq 3$.

4.4.1.3 Penalty Criteria

Tours constructed by the heuristics described above do usually not fulfill all temporal constraints. Capacity infeasibilities can occur due to applying DB for distributing remaining orders. In order to remove those infeasibilities we use a penalty approach in the improvement heuristics. The following parameters are used to penalize the various constraints when evaluating node n:

- π_1 - penalizes capacity infeasibilities (*i.e.*, an order is assigned to a vehicle v for which the number of sitting, lying patients or ones in wheelchairs exceed C_v^L, C_v^W or C_v^S values, respectively); the lateness, $d_{\nu n}$, associated with node n assigned to vehicle ν is set to $d_{\nu n} := \pi_1$ if the vehicle's capacity is exceeded,

- π_2 - used when an order is not in the vehicle's time windows (*i.e.*, an order is assigned to a vehicle which is "not yet" or "already not" available); $d_{\nu n}$ is set to $d_{\nu n} := \pi_2$

- π_3 - used to avoid that a patient needs to spend more than Δ_T minutes in a vehicle. If t denotes the time, a patient spends already in a vehicle, $d_{\nu n}$ is set to

$$d_{\nu n} := \pi_3 \max(0, t - \Delta_T) \quad .$$

4.4.2   Improvement Heuristics

The constructing phase of tours is followed by a *tour-improvement phase*. We have developed two different heuristic approaches:

- improving each tour T_v separately (sequencing heuristic; SH) ;

- improving tours by re-assigning orders to other vehicles (reassignment heuristic; RH).

The *sequencing* heuristic rearranges each tour, T_v, in such a way, that the sum of lateness and maximal lateness will decrease, if this is possible at all. Initially, T_v is characterized by S_{0v}^D. The procedure starts by identifying the node n_w, in this tour which has the maximum lateness and its associated pickup or delivery node, $\bar{n}_w$, *i.e.*, in each tour T_v we identify the node with maximal S_{0v}^D. The order o_w, which contains both the worst node and the corresponding pickup or delivery node, will be removed from T_v. These two nodes, which form o_w, will be inserted at possibly better positions in the same tour such that new characterizing values $S_v^D < S_{0v}^D$. This is done by the full enumeration scheme described in Subsection 4.4.1 for heuristic CONH2.

After all tours have been improved, one of the two following *reassignment* heuristics, IT1 and IT2, respectively, is applied to improve the tours.

4.4.2.1   Heuristic IT1

This heuristic starts with identifying the worst tour, T_w. The worst tour may either be identified by

- the total lateness, that is

$$w := \min \left\{ j : S_j^D > S_v^D, v = 1, \dots, j-1, \ j = 2, \dots, N^V \right\}$$

- considering both the maximal sum of lateness and the maximal lateness:

$$w := \min \left\{ j : (S_j^D > S_v^D) \wedge (D_j > D_v), \begin{array}{l} v = 1, \ldots, j - 1 \\ j = 2, \ldots, N^V \end{array} \right\}$$

Then, within tour $\mathcal{T}_w$ the worst order o_w is identified. Note that this is the order associated with the maximum lateness, d_{wn}. Order o_w (with its associated pickup and delivery nodes) will be removed from $\mathcal{T}_w$ and inserted into the tour of another vehicle.

A set of different strategies is provided to select the vehicle or tour, respectively, to which order o_w is reassigned:

- Insert order o_w, removed from $\mathcal{T}_w$, into the tour of that vehicle, which after checking all potential vehicles, has the smallest sum of lateness and smallest maximal lateness.

- Insert order o_w into that tour, for which this insertion leads to a minimal increase of the sum of lateness and the maximal lateness for the tour of this vehicle *w.r.t.* all other potential vehicles.

- Insert order o_w into tour $\mathcal{T}_b$, where $\mathcal{T}_b$ is the best existing tour, *i.e.*, tour with the minimal sum of lateness.

Note that it may happen that no vehicle exists which is able to take this order o_w according to the above criteria.

4.4.2.2 Heuristic IT2

This improvement heuristic, at the beginning, identifies the worst node, n_w^v, for each $\mathcal{T}_v$, $v = 1, \ldots, N^v$. The associated worst orders, o_w^v, corresponding to n_w^v will be removed from each $\mathcal{T}_v$ and collected in the set, named $\mathcal{O}^R$. The orders in the set $\mathcal{O}^R$ will be distributed in sequence to the vehicles, which would suffer less, *i.e.*, the sum of delays, S_v^D, are computed for each tour, $\mathcal{T}_v$, and the orders are added to the tour with minimal sum of lateness.

Notice, that it is best to use a combination of the *reassignment* heuristic RH and the *sequencing* heuristic SH. The computational experiments summarized in Appendix B.B.2.2 suggest to start improving tours with SH, then apply IT1, IT2 or a combination of them, and again apply SH.

4.4.2.3 Adding an Order to a Tour

To add an order i to an existing tour $\mathcal{T}_{0v}$, first, the sum S_{0v}^D of lateness in this tour, and the maximal lateness D_{0v} of $\mathcal{T}_{0v}$ are computed. The basic idea is to determine the best insertion points for these nodes (the pickup and the delivery nodes of i) by a complete enumeration procedure. In an outer loop, the pickup node n_1 is varied over all possible insertion points. Then, in an inner loop over all remaining points, the delivery node, n_2, is inserted (those nodes follow after the pickup point $w.r.t.$ the time, *i.e.*, $n_2 > n_1$). Let us denote this tour by $\mathcal{T}_v(i_1, i_2)$, the associated maximal lateness $D_v(i_1, i_2)$ and the sum of all deviations, $S_v^D(i_1, i_2)$. The best insertion points follows for all $k_1 = 1, \ldots, i_1 - 1$ and $k_2 = 1, \ldots, i_2 - 1$ as the couple

$$\{i_1^*, i_2^*\} := \min \left\{ \{k_1, k_2\} \left| \begin{array}{c} S_v^D(i_1, i_2) < S_v^D(k_1, k_2) \\ D_v(i_1, i_2) \leq D_v(k_1, k_2) + P_D \\ \mathcal{T}_v(i_1, i_2) \in \mathcal{T}_v^F \end{array} \right. \right\}, \quad (4.4.1)$$

where $\mathcal{T}_v^F$ is a set of all feasible tours for vehicle v. Note that the formula (4.4.1) contains the tuning parameter P_D which allows a certain additive increase of the maximal lateness (with respect to the minimal value found so far), while otherwise, we determine the minimum of the sum of all lateness over all feasible tours, *i.e.*, $\mathcal{T}_v(i_1, i_2) \in \mathcal{T}_v^F$. This tuning parameter, P_D, is one of the most sensitive parameters in the heuristic. Its value can be chosen in an experimental way, depending on D_v.

4.4.2.4   Termination Criteria

The improvement heuristics can be applied iteratively. For example, at first we apply IT1 and IT2, and when IT2 does not produce an improvement we can again apply IT1 and so on. How should we control the moves in our heuristics and when should we stop them? Let us first list some intuitive criteria and discuss whether they make sense:

- by number of iterations;

- by inspecting the standard deviation of the lateness in the tours;

- by inspecting the improvements in D_v and S_v^D;

- by comparing the values of D_v and S_v^D with some pre-given values;

- if $S_v^D = 0$ for all tour.

All these criteria might appear arbitrary and it is reasonable to discuss under which conditions they are useful. The number of iterations needed to converge may be different for different problems with different orders and vehicles. Certainly, D_v and S_v^D cannot become smaller than zero. On the other hand, if they are equal to zero, then this tour is optimal.

If this holds for all vehicles v then the overall solution, *i.e.*, the set of all tours, is optimal.

If all tours have similar values of S_v^D, *i.e.*, the standard deviation, σ_S, falls below a certain value, then not much can be achieved by moving an order from one vehicle to another one. This might be a good time to stop the iterative procedure.

Of course, if $S_v^D = 0$ for all vehicles we can stop because this is an optimal solution fulfilling all temporal constraints. This is the ideal stopping criteria although it probably will be become active only very seldom. Instead we can trace how S_v^D or $\bar{S}$ decrease. Let S and S_D be the sum of lateness and the sum of maximum lateness after reassigning an order from one vehicle to another one, and S^0 and S_D^0 the corresponding values before reassigning. There are two different schemes implemented which are applied if a new vehicle has been accepted in the cross-over routine:

1 If the conditions $\left|S - S^0\right| \leq \varepsilon$ and $\left|S_D - S_D^0\right| \leq \varepsilon$ are satisfied we stop indicating that there is no significant change.

2 If the inequalities $S^0 \leq S < 1.05 S^0$ hold, we accept this slight increase and stop.

It may happen that no stopping criteria becomes active and cycling occurs. A cycle appears, for instance, when the heuristic at first moves an order from vehicle v_1 to v_2 ($v_1 \longrightarrow v_2$) and the next time back, from vehicle v_2 to v_1 ($v_2 \longrightarrow v_1$). A cycle more complicated to detect looks as follows: $v_1 \longrightarrow v_2$, $v_3 \longrightarrow v_4$, $v_2 \longrightarrow v_1$, $v_4 \longrightarrow v_3$. We check for both types of cycles and change one time from IT1 to IT2. If this change has already been applied and cycling is identified, the reassignment heuristic stops.

EXAMPLE 4.10 *In this example we illustrate the final solutions using IT1, IT2 and CONH1 and CONH2 with DB, DU for distributing remaining orders. Also we combine IT1 and IT2. For this particular example CONH1 produces the same result in combination with improvement heuristics as CONH2. So, further we do not indicate which one is used. In combination of IT1 and IT2 both heuristics are applied consecutively.*

DU and IT1 $\left(\max\limits_{v=1,2} D_v = 0.26, \sum_{v=1}^{2} S_v^D = 0.43 \right)$:

v	D_v	S_v^D	T_v
1	0.26	0.28	$i_{o_1} \rightarrow i_{o_4} \rightarrow i_{o_5} \rightarrow j_{o_1} \rightarrow j_{o_4} \rightarrow j_{o_5}$
2	0.15	0.15	$i_{o_3} \rightarrow i_{o_2} \rightarrow j_{o_3} \rightarrow j_{o_2}$

DU and IT2 $\left(\max\limits_{v=1,2} D_v = 0.26, \sum_{v=1}^{2} S_v^D = 0.42 \right)$:

v	D_v	S_v^D	T_v
1	0.26	0.42	$i_{o_1} \to i_{o_4} \to i_{o_3} \to j_{o_1} \to j_{o_3} \to j_{o_4}$
2	0.00	0.00	$i_{o_2} \to i_{o_5} \to j_{o_5} \to j_{o_2}$

DU and combination of IT1 and IT2 $\left(\max_{v=1,2} D_v = 0.26, \sum_{v=1}^{2} S_v^D = 0.42 \right)$:

v	D_v	S_v^D	T_v
1	0.26	0.42	$i_{o_1} \to i_{o_4} \to i_{o_3} \to j_{o_1} \to j_{o_3} \to j_{o_4}$
2	0	0	$i_{o_2} \to i_{o_5} \to j_{o_5} \to j_{o_2}$

DB and IT1 $\left(\max_{v=1,2} D_v = 0.26, \sum_{v=1}^{2} S_v^D = 0.41 \right)$:

v	D_v	S_v^D	T_v
1	0.26	0.29	$i_{o_1} \to i_{o_3} \to i_{o_5} \to j_{o_1} \to j_{o_3} \to j_{o_5}$
2	0.12	0.12	$i_{o_4} \to i_{o_2} \to j_{o_4} \to j_{o_2}$

DB and IT2 $\left(\max_{v=1,2} D_v = 0.26, \sum_{v=1}^{2} S_v^D = 1.00 \right)$:

v	D_v	S_v^D	T_v
1	0.26	1.00	$i_{o_1} \to i_{o_4} \to i_{o_3} \to i_{o_5} \to j_{o_3} \to j_{o_1} \to j_{o_4} \to j_{o_5}$
2	0.00	0.00	$i_{o_2} \to j_{o_2}$

DB and combination of IT1 and IT2 $\left(\max_{v=1,2} D_v = 0.26, \sum_{v=1}^{2} S_v^D = 0.41 \right)$:

v	D_v	S_v^D	T_v
1	0.26	0.29	$i_{o_1} \to i_{o_3} \to i_{o_5} \to j_{o_1} \to j_{o_3} \to j_{o_5}$
2	0.12	0.12	$i_{o_4} \to i_{o_2} \to j_{o_4} \to j_{o_2}$

*This simple example is too small to see the efficiency of combining **IT1** and **IT2**, but in offline cases the combination of these two heuristic can improve the solution significantly.*

4.4.3　Simulated Annealing and Vehicle Routing

Simulated Annealing (SA) belongs to the class of metaheuristics for solving optimization problems using a simulation-based approach. Although SA can be applied to continuous optimization problems [87], it is frequently used for finding good, not necessarily optimal solutions to a wide variety of combinatorial optimization problems. The original idea of SA stems from statistical mechanics and Monte Carlo techniques. It can be seen as a generalization of a Monte Carlo method for examining the equations of state and frozen states of n-body systems [59], especially the slow cooling of metals. The Monte Carlo method is coupled to a probability acceptance criterion, which allows to escape from local optima. While often the global optimum is reached, the method neither can guarantee this, nor can it provide a safe bound on a minimization problem. However, by finding feasible points it can provide an upper bound for minimization problems.

As other metaheuristics, SA has been applied to vehicle routing problems (*cf.*, [62],[16]). In this section we present our own formulation of the problem to be set up for an SA routine. This enables us with little programming overhead to treat larger problems and to find (hopefully) good feasible tours. Especially, we use the SA for offline optimization scenarios with 200 to 300 orders.

4.4.3.1 Monte Carlo Methods and Simulated Annealing

In its simplest form, Monte Carlo simulation samples the possible states of a system by randomly choosing evaluation points. The ensemble of randomly chosen points in the search space provides some information about this space. This procedure is useful in some problems, *e.g.*, finding the area of regions bounded by a complicated curve. Unlike choosing those points randomly, in SA a new point in search space is sampled by making a slight change to the current point, *i.e.*, a new point is selected in the vicinity or neighborhood of the current point. The choice of the neighborhood depends strongly on the problem. To illustrate the neighborhood concept, consider a new orientation of the helium atoms created by making a random, small change to the coordinates of each atom. If the energy of this new orientation is less than that of the old one, this orientation is added to the ensemble. If the energy rises, a *Boltzmann acceptance criteria* is used. If the energy rise is small enough to fulfill a probabilistic rule involving a term such as $e^{-\Delta E/T}$ with energy change ΔE, and temperature T, the new orientation is added to the ensemble. Conversely, if the energy rise ΔE is too large, the new orientation is rejected and the old orientation is again added to the ensemble. What is unique about this Boltzmann acceptance probability is that the temperature of the system must be used.

In 1983, Kirkpatrick and colleagues [52] proposed a method using a Metropolis Monte Carlo simulation to find the lowest energy (most stable) orientation of a physical system. Their method is based upon the procedure used to make the strongest possible glass. This procedure heats the glass to a high temperature so that the glass is a liquid and the atoms can move relatively freely. The temperature of the glass is slowly lowered so that at each temperature the atoms can move enough to begin adopting the most stable orientation. If the glass is cooled slowly enough, the atoms are able to "relax" into the most stable orientation. This slow cooling process is known as *annealing* and gives the name to this method *simulated annealing*.

4.4.3.2   Optimization Problems and Simulated Annealing

Let us now consider how SA can be applied to optimization problems. Combinatorial optimization problems are often represented by graphs. Let us therefore consider a graph with an energy E assigned to each node. The energy is to be minimized, *i.e.*, the energy plays the role of the objective function, $f(\mathbf{x})$, in the optimization. In the parlance of SA, the nodes are called 'states', the arcs represent 'moves' or 'transitions' from one state to a neighboring state, and the energy is sometimes called 'cost'. Table 4.4 compares and relates the language used in both disciplines:

Table 4.4: Analogies between the physics and optimization

Physics	Optimization
energy	objective function
states, configuration	nodes of a graph, or the decision variables, $\mathbf{x}$
moves, transition	arcs of a graph
false solution	local solution
ground state	global minimum

The analogy and the generalization of this Monte Carlo approach to combinatorial problems is straightforward ([15], [52]) for some of the objects involved. The current state of the thermodynamic system is analogous to the current solution to the combinatorial problem, the energy function is analogous to the objective function, and the ground state is analogous to the global minimum. The major difficulty and art in implementing the SA algorithm is that there is no obvious analogy for the temperature T with respect to a free parameter in the combinatorial problem. Furthermore, avoidance of entrainment in local minima (quenching) is dependent on the "annealing schedule", the choice of the initial temperature, T_0, the choice of the number of iterations, N_T, to be performed at each temperature, and the choice of temperature decrement at each step as cooling proceeds described by either ΔT or δ.

With a randomly chosen initial state, S_0, the SA algorithm then works as follows:

initialize *state* $S = S_0$
repeat the outer loop (until done)
{

    $T = $ new temperature, T', repeat (until inner-loop criterion)
    { generate *new state* $S' = $ random neighbor(S)
       compute the *energy difference* $\Delta E = E(S') - E(S)$
       re-set *state* $S = \begin{cases} S', \text{ if } \Delta E \leq 0 \\ S' \text{ with } probability \ e^{-\Delta E/T}, \text{ if } \Delta E > 0 \end{cases}$

    }

}

The temperature T is *the* control parameter of the algorithm. It is always decreased gradually as the annealing proceeds, but the choice of an optimal control of T is a very subtle issue. While annealing works well on a wide variety of practical problems, like other metaheuristics it always requires considerable fine-tuning depending on the optimization problem at hand:

- What is the best initial temperature?

- How fast should be done cooling?

- How long should the simulation be run at each temperature, *i.e.*, how many time the inner loop should be execute?

- When should the outer loop be terminated?

The initial temperature depends on the expected value of the energy, *i.e.*, objective function. The answer to the third question depends upon the maximum size of the Monte Carlo step at each temperature. While a pure Metropolis Monte Carlo simulation attempts to reproduce the correct Boltzmann distribution at a given temperature, the inner-loop of SA optimization only needs to be run long enough to explore the regions of the search space that should be reasonably populated. This allows for a reduction in the number of Monte Carlo steps at each temperature, but the balance between the maximum step size and the number of Monte Carlo steps is often difficult to achieve, and depends very much on the characteristics of the search space or energy landscape.

4.4.3.3 Improving Tours by Simulated Annealing

In order to improve tours or the driving schedule of the whole vehicle fleet we have to develop a neighborhood concept, a transition mechanism and a cooling scheme. We investigated two different SA implementations:

- improving only each tour for each vehicle separately, and

- reassigning orders from one tour to another one.

Improving tours individually, *i.e.*, single-vehicle tour improvement, produces usually an improvement by 20 to 30% compared to using heuristics alone. Our implementation of SA applied to reassigning orders was not successful.

Let us start by adjusting SA variant for improving an individual tour. If we neglect the temporal and capacity constraints for the moment, such a tour can be understood as a tour in the traveling salesman problem

(TSP) with a different objective function. However, this objective function is not the sum of terms which only depend on the distance between adjacent nodes but an objective function depending on the whole tour characterized by the sum of lateness and maximal lateness. Nevertheless, this analogy with the TSP helps us to define the neighborhood concept or to define appropriate transitions.

In TSP-like problems one often just reverses the direction between two nodes. Therefore, we will call this operation *swap*. It just depends on two numbers n_1 and n_2, $n_2 \neq n_1$, $1 \leq n_1, n_2 \leq N$ and transforms a given tour T with N nodes $\nu_1, \nu_2, \ldots, \nu_N$

$$T := \{\nu_1, \nu_2, \ldots, \nu_{n_1-1}, \nu_{n_1}, \nu_{n_1+1}, \ldots, \nu_{n_2-1}, \nu_{n_2}, \nu_{n_2+1}, \ldots, \nu_N\}$$

into

$$T^R := \{\nu_1, \nu_2, \ldots, \nu_{n_1-1}, \nu_{n_2}, \nu_{n_1+1}, \ldots, \nu_{n_2-1}, \nu_{n_1}, \nu_{n_2+1}, \ldots, \nu_N\}.$$

Note that the only requirement is $n_2 \neq n_1$. In the example we have $n_2 > n_1$ but this has been chosen only for demonstration purposes.

Alternative to this operation *swap* we may also wish to shift whole tour segment. This operation is called *shift* and is characterized it by three numbers $n_1 < n_2$, and n_3; with $1 \leq n_1, n_2, n_3 \leq N$ and another condition described below. Let us demonstrate this operation by a small tour consisting of 10 nodes and $n_1 = 3$, $n_2 = 6$ and $n_3 = 4$. At first we extract the node segment starting at node 3 and ending at node 6 from the original tour

$$T := \{\nu_1, \nu_2, \nu_3, \nu_4, \nu_5, \nu_6, \nu_7, \nu_8, \nu_9, \nu_{10}\}$$

which gives

$$T' := \{\nu_1, \nu_2, \nu_7, \nu_8, \nu_9, \nu_{10}\} \quad , \quad T_e := \{\nu_3, \nu_4, \nu_5, \nu_6\}$$

where T_e is the extracted path segment. Note that T_e contains

$$N(T_e) := n_2 - n_1 + 1$$

nodes. Now we insert the node segment T_e after position n_3 of T' which gives

$$T^S := \{\nu_1, \nu_2, \nu_7, \nu_8, \nu_3, \nu_4, \nu_5, \nu_6, \nu_9, \nu_{10}\} \quad .$$

From this small example we can derive the necessary condition

$$N - n_3 + 1 \geq N(T_e) \Leftrightarrow n_3 \leq N - N(T_e) + 1$$

for obtaining a feasible shift.

We can now define how often the *swap* and the *shift* operations should be applied. We want to use these operations randomly. Thus, we just need to define the frequency of one operation with respect to another one. Let ρ, $0 < \rho < 1$, define the probability that we choose the *shift* operation. If we let a uniform random generator decide which operation to choose, we obtain the following limit:

$$\lim_{k \to \infty} \frac{n^S}{n^R} = \frac{\rho}{1 - \rho} \quad .$$

As described in Section 5.2.6, $\rho = 0.0125$ turns out to be a good value.

Let us now concentrate on the cooling scheme. We start with an initial cooling temperature T_0. If z is the objective function value associated with a tour before applying SA, our numerical experiments in Section 5.2.6 show that $T_0 = 0.1z$ is a good value. For each sequence of n successful transitions, the cooling temperature is decreased according to the proportional cooling scheme

$$T_k = \delta T_{k-1} \quad , \quad k = 1, \ldots, K$$

with $\delta = 0.9$ established in Section 5.2.6.

Treating Constraints. Constraints are enforced by penalizing violations as this gives maximal flexibility to incorporate all the features present in the real world problem. As was mentioned in Section 4.4.1.3, penalty parameters π_1 and π_2 are applied to capacity and time window constraints. Violations of the time windows related to the availability of vehicles are penalized by the penalty parameter π_3.

Objective Function. The objective function is composed of the total lateness, S_v and the maximum lateness, D_v. We might choose, for instance, $z = S_v + W_D D_v$ with $W_D = 15$. Other test runs with different values of W_D gave worse results. The values S_v and D_v contain penalty contributions if a tour violates a temporal or capacity constraint.

Termination Criteria. Here we might choose either the number of iterations, *i.e.*, how often the temperature has been decreased, the temperature itself, or the number of successful moves. In our numerical experiments we stopped after $N_T = 100$ temperature decreases.

Applications Variants. There are many degrees of freedom in applying SA to our problem. We use it either after the first improvement of the constructed tours (IFSA1= 1), then start the cross-over heuristics, and then apply it again to the each individual tour (IFSA2= 1), or alternatively (IFSA1= 0, IFSA2= 1), we apply it only after the cross-over heuristic finished. The latter worked significantly better. This can be understood as follows. The better the solutions of the individual tours, the more difficult is it for the cross-over heuristic to obtain successful moves and to reassign an order to a different vehicle.

We also developed a SA scheme for cross-over and inter-tour optimization, but this approach turned out not to be successful. Almost all attempts to reassign an order to another vehicle did not improve the objective function used in SA. Thus, we gave up on this and used SA only for intra-tour optimization.

4.5.   Summary

In this chapter we have developed several approaches to solve VRP-PDTWs in hospital transportation. These include exact optimization algorithms (MILP and column enumeration) and heuristic methods. The exact methods are designed to solve online problems of small or modest size. While the MILP approach (Section 4.3.1) often requires a long time to prove optimality, the column enumeration approach (CEA, Section 4.3.3) allows to compute optimal solutions very fast. This approach is based on decomposing the problem into a master problem (a set partitioning problem controlling the assignment of orders to vehicles) and a set of subproblems (intra-tour problem, routing and scheduling). To solve the routing and scheduling problem exactly we have developed a branch-and-bound method which prunes nodes if they are value-dominated by previous found solutions or if they are infeasible *w.r.t.* the capacity or temporal constraints. This approach works fast for up to 10 orders per vehicle. This branch-and-bound method is suitable to solve any kind of sequencing-scheduling problem involving cumulative objective functions and constraints, which can be evaluated sequentially.

To solve intra-tour problems containing more than 10 orders we have developed various construction and improvement heuristics (Section 4.4). Especially, to treat larger offline problems with several hundred orders we developed a heuristics to reassign orders to other vehicles, and improve existing tours by simulated annealing (Section 4.4.3).

Chapter 5

VEHICLE ROUTING PROBLEMS IN HOSPITAL TRANSPORTATION. II APPLICATIONS AND CASE STUDIES

In Chapter 4 algorithms and heuristic methods have been developed to solve the vehicle routing problem arising in hospital transportation. Here we apply them to the real world problem described in Section 5.1. Real world data enable us to test the models and the heuristics (see Section 5.2).

5.1. OptiTrans

Here we briefly review only those aspects of the hospital project "Opti-Trans" with the *Klinikum des Saarlandes in Homburg* which are relevant to the online optimization requirements, because this joint project between the ITWM and the *Klinikum des Saarlandes in Homburg* has been described earlier in great detail [22].

Homburg has a major hospital campus consisting of many buildings spread over 300 ha (see Fig. 5.1.1). The transportation inside the campus is served by a vehicle fleet. One may need to use a vehicle, because some buildings are located very far from each other. Further, the physical state of some patients may also require transportation by a vehicle (passenger transport). Moreover, some passengers need to lie in bed while being transported, others can sit, others can only sit in a wheel chair, some need assistance, *i.e.*, they are accompanied. A few need a special car which carries them in their bed. And again, some others are infectious persons requiring that the car is completely cleaned after a tour.

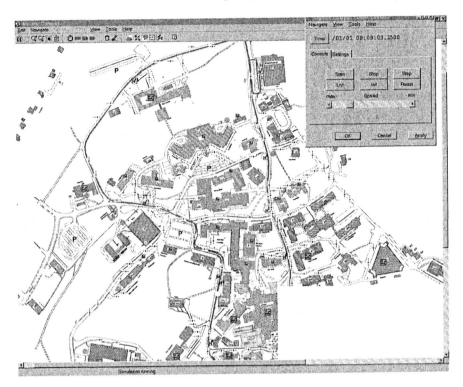

Figure 5.1.1. A map showing the hospital campus of the Homburg University

In addition, various objects (medical goods, laundry, food, documents, etc.) have to be transported between buildings every day (the buildings will be called as "pickup location" and "delivery location").

Usually hospitals run their own vehicle fleet to cover their transportation needs. There are 13 vehicles in the Homburg hospital. There are two hospital employees responsible for one vehicle. Each vehicle provides space for two lying and five sitting passengers. The main daily operating time is from 7am to 5pm, but as shown in Fig. 5.1.2 there are also orders at other times.

The dispatcher in the central office at first collects the data characterizing orders, and then passes them to the drivers (Fig. 5.1.3). Orders are characterized by their name, pickup and delivery location, required or preferred pickup and delivery times, number and type of passengers or objects to be transported, accompanying personnel to be transported, service times for picking-up and delivering a patient. Special orders are those involving infectious passengers or goods, or those requiring a doctor accompanying the passengers. The dispatcher has to assign those

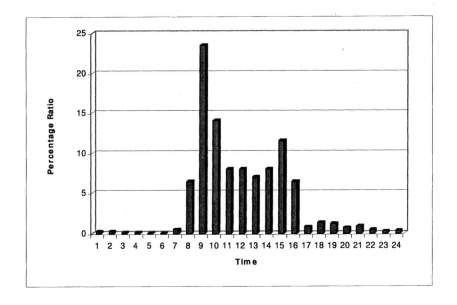

Figure 5.1.2. Time distribution of the orders over the day

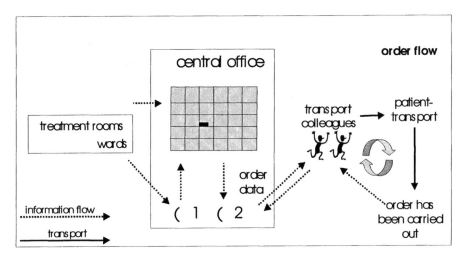

Figure 5.1.3. Information flow in hospitals

orders to the drivers and vehicles, keeping in mind that waiting and driving times as well as empty drives have to be minimized while passenger satisfaction has to be maximized (see Fig. 5.1.4).

Each day, about 200 to 300 transport orders have to be served by this fleet, most of them between 7am and 5pm. The life of the dispatcher becomes more complicated because many orders arrive at short notice. Every hour about 20 to 30 new orders appear and need to be assigned and scheduled to the available vehicles requiring that tours can be modified and extended. Typically, one or two passengers are already in a vehicle, when a new order, *i.e.*, pickup location is assigned to a vehicle. The orders associated with the passengers already loaded to the vehicle are called *pre-assigned* or *active* orders (*i.e.*, not all old orders are fulfilled when new orders arrive: some patients are still on the way to their destinations).

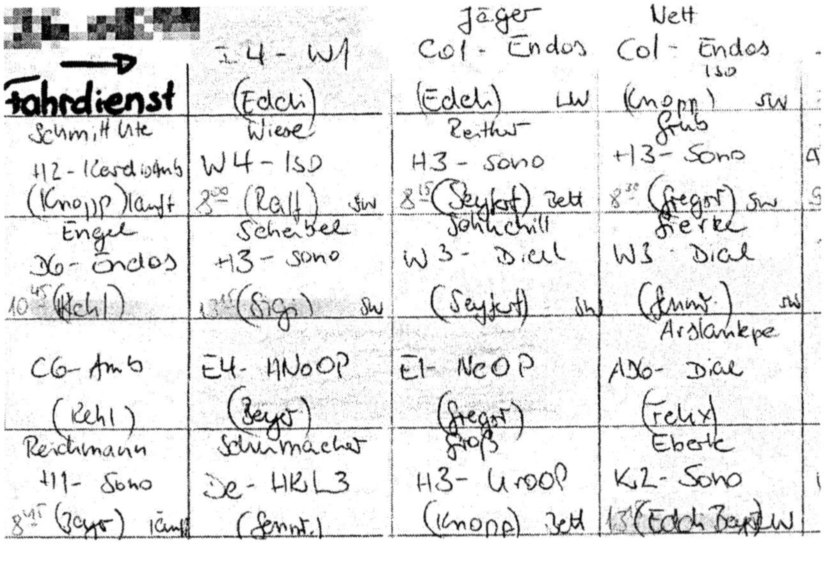

Figure 5.1.4. Scheduling of orders to the vehicles that done by hands

Orders are allocated to the vehicles according to their availability. If there are many orders with the same target time it is important to ensure that the vehicles are exploited appropriately. Some vehicles may not be available at all or only partially available during the day, *i.e.*, each vehicle has its own time window.

Homburg's university hospital is one of the very few hospitals which has already implemented an IT-based transport system. This provides us with a database large enough to use mathematical optimization and optimal routing to allocate orders to vehicles. The optimization model helps to allocate them in such a way that all of them are fulfilled in time, or at least with minimal delay and the capacity of each vehicle is not exceeded.

Knowing all orders for a day in advance would lead us to an offline optimization problem in which a tour starts and ends at the depot. However, the reality of this hospital transportation needs leads us to an online optimization situation in which we find a vehicle in a given location and it stops at the last order served. Of course, in order to evaluate the efficiency of our online optimization strategies we need to be able to compare them to the optimal solution of the offline problem. When we do online optimization it may happen that the patient is already in the vehicle. In this case, for this *active* order we have to consider additional input information, *i.e.*, the number of this vehicle and how long this patient is already in the vehicle. For offline optimization these data are not required.

Our online approach solves the online version of the problem but uses an exact optimization method to do so, *i.e.*, to solve this online version. For solving online optimization scenarios it suffices to solve problems which add at most six orders to a vehicle; sometimes there are already two passengers in the vehicle, *i.e.*, two pre-assigned orders present. Thus, in total, eight orders need to be assigned and scheduled. In peak hours there are about 20 to 30 orders, but the dispatcher assigns new orders and re-schedules existing tours every few minutes; actually, he does this immediately, when a new order arrives. Thus, the problem size remains small.

5.2.   Numerical Tests

5.2.1   Input data

Four ASCII data files contain the required input data:

1 a list of orders each characterized by several parameters;

2 a list of vehicles with time windows indicating when each is available, and the current time;

3 a distance matrix (usually non-symmetric);

4 control files assigning values to the various control parameters.

The content of these files is described in detail in Appendix B.B.1.1.

5.2.2   Offline and Online Versions

Both, the models and the heuristics, are tested for online versions. Moreover, we compare the solution, produced by the heuristics for the offline- and online- versions. For the offline version the current time is not relevant. All orders are known in advance and all vehicles are located at their depot. For the online version one or several orders could be re-planned. At the current time, T_*, all vehicles may be positioned at any location. After re-planning, the vehicle remains where it served the last order. It may also happen that a vehicle has already picked up a passenger, and is just driving from this pickup point to another location when re-planning starts.

In this section several experiments under various conditions are presented:

1 intra-tour optimization (no assignment problem involved, single vehicle experiments);

2 an offline optimization scenario covering a full day (200 to 300 orders); and

3 online optimization scenarios including pre-assigned orders.

5.2.3 MILP Model

The MILP model involves too many binary variables to apply it to the offline case covering a whole day. Thus, in this section, we present only experiments for

1 intra-tour optimization (no assignment problem involved, single-vehicle experiments); and

2 online optimization scenarios for the whole problem (assignment plus routing and scheduling) including pre-assigned orders.

In the sequel we use the following notations:

- β is the number of binary variables;

- r is the number of requests (orders), involved in the model;

- N is the number of nodes in the branch-and-bound tree of the MILP solver.

The cases discussed in this section are named in sequence, M01, ..., M10 and their detailed tabular results are collected in Appendix B.B.2.1.

5.2.3.1 Intra-Tour Optimization (Single-Vehicle Cases)

In this section we analyze an offline data set to illustrate how strongly the performance of the MILP approach depends on the data and structure of the problem. From a given set of orders with overlapping pickup requests we generated several test examples which only differ by the number of orders. The input data are shown in Table 5.1.

Table 5.1: The input data of the single-vehicle experiment

i	order	pickup	delivery	type	flag1	flag2	serv1	serv2
1	A150826	7.50000 ,	8.00000 ,	1 ,	0 ,	0 ,	4 ,	2
2	A150752	7.50000 ,	8.00000 ,	2 ,	0 ,	0 ,	4 ,	5
3	A150869	8.00000 ,	8.50000 ,	2 ,	0 ,	0 ,	0 ,	0
4	A150890	8.00000 ,	8.50000 ,	2 ,	0 ,	0 ,	14 ,	1
5	A150922	8.58333 ,	9.08333 ,	1 ,	0 ,	0 ,	12 ,	8
6	A150920	9.00000 ,	9.50000 ,	1 ,	0 ,	1 ,	15 ,	5
7	A150939	9.50000 ,	10.00000 ,	2 ,	0 ,	0 ,	0 ,	0
8	A150946	9.60000 ,	10.10000 ,	1 ,	0 ,	0 ,	12 ,	7
9	A150965	9.98333 ,	10.48333 ,	1 ,	0 ,	0 ,	2 ,	3
10	A150970	10.25000 ,	10.75000 ,	2 ,	0 ,	0 ,	0 ,	0
11	A150977	10.25000 ,	10.75000 ,	1 ,	0 ,	0 ,	16 ,	8
12	A150749	11.00000 ,	11.50000 ,	2 ,	0 ,	0 ,	0 ,	0
13	A151000	11.25000 ,	11.75000 ,	1 ,	0 ,	0 ,	9 ,	9
14	A150933	11.50000 ,	12.00000 ,	2 ,	0 ,	0 ,	14 ,	3
15	A151038	12.51667 ,	13.01667 ,	1 ,	0 ,	0 ,	18 ,	6
16	A151022	13.25000 ,	13.75000 ,	1 ,	0 ,	0 ,	9 ,	6

```
type : 1 (lying in bed passenger), 2 (sitting passenger)
flag1: indicate, whether the patient is infectious
flag2: number of accompanying personnel
serv1: service time (in min.) required to pickup the patient
serv2: service time (in min.) required to deliver the patient
```

Especially difficult to assign are the pairs of orders ({1,2},{3,4},{10,11})
with identical pickup times, because the usually lead to situations in
which a vehicle arrives significantly late. Table 5.2 shows the results
of the Xpress-MP MILP solver when solving the offline scenarios M01,
M02, M03, and M04:

Table 5.2: The results of the scenarios M01, M02, M03, and M04

name	r	β	N	time/s	z
M01	4	65	3,184	10	0.179957
M02	5	97	16,711	64	0.180045
M03	6	125	39,720	271	0.194901
M04	7	149	384,649	3,360	0.194997

All problems have been solved without any special tuning of the Xpress-
MP MILP solver (Release 13.26) or special tightening of the model. The

objective function values are optimal up to a percentage cut of 1%. Already for seven orders (14 nodes in the tour) the problem cannot be solved easily and in reasonable time. In order to analyze somewhat larger scenarios we apply the arc elimination techniques described in Section 4.3.1.7, and also the cuts (4.3.19) and (4.3.20). Here we report on two experiments using this approach. Case 1 covers 12 orders {1,2,3,4,5,6,11,12,13,14,15,16} while case 2 covers all 16 orders. The results are shown in Table 5.3:

Table 5.3: The results of the scenarios M05 and M06

name	r	Δ^B	Δ^F	R^{A_2}	R^B	β	nodes	time/s	z
M05	12	1.00	1.00			149	45496	188	0.265996
M06	16	0.75	0.75	0.2	0.2	241	6,468,076	117,836	0.283891

Note that R^{A_2} and R^B have been set to 0.2 because the maximal lateness of the tour obtained by the heuristics was somewhat smaller than this value. These two examples show how the solution time and number of nodes required to prove optimality grow as functions of the number of binary variables. Only four more orders lead to approximately additional 100 binary variables. The running time increases, however, by a factor of 626.

What is said in the previous paragraph might give an impression that the MILP model, even in the single-vehicle case, is not useful. This is definitely not so. Below we summarize two experiments with 13 to 16 orders but this time we supported the branch-and-bound process with extra knowledge derived from the solution of the heuristic. For such feasible solutions we know the sum of lateness which is exploited to set the upper bound, S_L, used in (4.3.42). Table 5.4 shows the results of the exact optima (no percentage cut) of the MILP model. Case M09 includes all 16 orders while M07 and M08 do not include the orders {2,11,13} and {2,13}, respectively.

Table 5.4: The results of the exact optima based on the MILP model

case	r	Δ^B	Δ^F	R^{A_2}	R^B	S_L	β	nodes	time/s	z
M07	13	1.00	1.00			0.16	199	487	3	0.041216
M08	14	1.00	1.00			0.10	227	834	3	0.074163
M09	16	0.70	0.70	0.2	0.2	0.30	237	247,268	1408	0.283887

Note that M09 produces the same solution and detailed tour as M06, but in a much shorter time. Thus the detailed tour is not displayed in the appendix.

The small examples in this section show that the performance depends critically on the structure of the problem, less on the number of binary variables. The objective function values correspond to those obtained by the heuristic approach. The main difference between the *easy* cases M07 and M08, on the one hand, and the *difficult* case M09, one the other hand, is that in the cases M07 and M08 the objective function value is almost zero, while in the latter case it is not. Therefore, in M09 one observes the general problem intrinsic to the problem when using a MILP model: the lower bound increases only very slowly. This is almost equivalent to the fact that feasible integer solutions are only found deep in the tree. If the tree is large it can take a very long time to reach those nodes deep in the tree.

We conclude this set of numerical experiments by summarizing that the MILP model can optimally solve single-vehicle problems with a smaller number of orders, but in each specific case it depends strongly on the input data and fine tuning. Thus, it might not be easy to implement this approach into a decision support tool which should work similar in all situations.

5.2.3.2   An Online Case Including Pre-assigned Orders

Consider the set of orders presented in Table 5.5 containing the information about the availability of the vehicle from Table 5.6. Here $N^O = 14$, $N^V = 12$. Let the current time be $T_* = 14.128$ and $C_v^L = 2$, $C_v^S = 5$. Notice, that some orders have pickup times earlier than the current time (this means that these orders will be definitely delayed). Some vehicles have already patients inside, for instance:

- vehicle 2 has one lying patient (order $A25525$),

- vehicle 3 has one lying patient (order $A25532$),

- vehicle 4 has two sitting patients (orders $A25524$ and $A25504$),

- vehicle 5 has two sitting patients (orders $A25501$ and $A25531$),

- vehicle 6 has one sitting patient (order $A25508$),

- vehicle 7 has one sitting patient (order $A25441$).

The last column of Table 5.5 indicates how many minutes, Δ_*, the passenger is already in the vehicle.

Table 5.5: Input data for an online case study

order	T^E	type	car	Δ_*
A25524	13.500	sitting	4	5
A25501	13.750	sitting	5	5
A25504	13.750	sitting	4	20
A25528	13.766	lying	0	0
A25525	14.000	lying	2	5
A25531	13.850	sitting	5	5
A25532	13.888	lying	3	5
A25527	14.000	lying	0	0
A25508	14.000	sitting	6	5
A25441	14.000	sitting	7	5
A25364	14.450	lying	0	0
A25530	14.000	sitting	0	0
A25533	14.500	lying	0	0
A25534	14.250	lying	0	0

While all vehicles are available from the current time, $T_* = 14.128$, they differ in the latest time of their availability, as shown in the following Table 5.6.

Table 5.6: The input data of the single-vehicle experiment

```
-----------------------------------------------------------------

v :   1    2    3    4    5    6    7    8    9   10   11   12

-----------------------------------------------------------------

    24.0 14.2 14.2 15.2 15.7 16.0 16.0 16.2 16.2 16.2 16.7 16.7

-----------------------------------------------------------------
```

The best solution obtained using the MILP model is shown in Table 5.7.

Table 5.7: The best solutions obtained by the MILP model

v	node	TE	tA	tL	rA2	rB	S L
1	25528		14.308	14.266	–	0.042389	0 0
1	A25528	13.766	14.143		0.377089	–	0 1
1	DD1		14.458		–	–	0 0
2	A25525	14.000	*		–	–	0 1
2	25525		14.203	14.500	–	–	0 0
2	DD2		14.377		–	–	0 0

3	A25532	13.888	*		–	–	0	1
3	A25527	14.000	14.168		0.168056	–	0	2
3	25532		14.371	14.388	–	–	0	1
3	25527		14.421	14.500	–	–	0	0
3	DD3		14.484		–	–	0	0
4	A25504	13.750	*		–	–	2	0
4	A25524	13.500	*		–	–	1	0
4	25504		14.190	14.250	–	–	1	0
4	25524		14.319	14.260	–	0.059423	0	0
4	A25533	14.500	14.500		–	–	0	1
4	25533		15.000	15.000	–	–	0	0
4	DD4		15.160		–	–	0	0
5	A25501	13.750	*		–	–	1	0
5	A25531	13.850	*		–	–	2	0
5	25501		14.190	14.250	–	–	1	0
5	25531		14.321	14.350	–	–	0	0
5	DD5		14.425		–	–	0	0
6	A25508	14.000	*		–	–	1	0
6	25508		14.185	14.500	–	–	0	0
6	DD6		14.341		–	–	0	0
7	A25441	14.000	*		–	–	1	0
7	A25534	14.250	14.250		–	–	1	1
7	25441		14.408	14.500	–	–	0	1
7	25534		14.750	14.750	–	–	0	0
7	DD7		14.839		–	–	0	0
9	A25364	14.450	14.450		–	–	0	1
9	25364		15.026	14.500	–	0.525900	0	0
9	DD9		15.258		–	–	0	0
12	A25530	14.000	14.144		0.143923	–	1	0
12	25530		14.238	14.500	–	–	0	0
12	DD12		14.367		–	–	0	0

Each line of the Table 5.7 contains the following data (from the left to the right):

1 the number of the vehicle, v;

2 the name of the node (a node starting with A is a pickup node);

3 the earliest permissible arrival time, T_i^E, at a pickup node (this is the desired target time);

4 the real arrival time, t_n^A, in the solution (an asterisk indicates that the passengers assigned to these vehicles were already loaded to the car);

5 the latest permissible time, T_j^L, (target time for delivery nodes);

6 the lateness, r_i^{A2}, for pickup nodes (a hyphen indicates there was none at all);

7 the lateness, r_j^B, for delivery nodes (a hyphen indicates there was none at all);

8 the number, p_v^S, of sitting passengers in the vehicle; and

9 the number, p_v^L, of lying passengers in the vehicle.

The objective function was (4.3.1) evaluated with $C_{nvt}^P = 1$, and all other components of (4.3.1) were set to zero, *i.e.*, $z := c^T = c^P$ contained only the penalty cost term for earliness and lateness. The objective function value after 500,000 nodes (or 16 hours CPU time) was $z = 1.316780$; the best lower bound was $z^- = 1.202968$ which corresponds to an integrality gap of $\Delta = 8.6\%$. This values were already obtained after 4000 nodes, or 4 minutes. In this example, during 16 hours of CPU time the lower bound increased by less than 10^{-6}. Thus, it may take very long to prove optimality with the MILP solver in this case. Using the column enumeration approach in Section 5.2.4, this value is proven to be optimal.

Note that in one case vehicle 3 transports two lying passenger simultaneously. Strictly speaking, the vehicle has capacity for two lying passengers but the dispatcher usually tries to avoid this situation. To avoid this situation, in another run c^L was included in the objective function with $C_{nv}^{LL} = C_{nv}^{LS} = 0.001$.

Table 5.8: Optimal tours with no more than one lying passenger

v	node	TE	tA	tL	rA2	rB	S	L
1	A25528	13.766	14.143		0.377089	–	0	1
1	25528		14.308	14.266	–	0.042389	0	0
1	DD1		14.458		–	–	0	0
2	A25525	14.000	*		–	–	0	1
2	25525		14.203	14.500	–	–	0	0
2	A25364	14.450	14.450		–	–	0	1

2	25364		15.026	14.500	–	0.525900	0	0
2	DD2		15.258		–	–	0	0
3	A25532	13.888	*		–	–	0	1
3	25532		14.153	14.388	–	–	0	0
3	A25534	14.250	14.250		–	–	0	1
3	25534		14.424	14.750	–	–	0	0
3	DD3		14.545		–	–	0	0
4	A25504	13.750	*		–	–	1	0
4	A25524	13.500	*		–	–	2	0
4	25504		14.190	14.250	–	–	1	0
4	25524		14.319	14.260	–	0.059423	0	0
4	A25533	14.500	14.500		–	–	0	1
4	25533		15.000	15.000	–	–	0	0
4	DD4		15.160		–	–	0	0
5	A25501	13.750	*		–	–	1	0
5	A25531	13.850	*		–	–	2	0
5	25501		14.190	14.250	–	–	1	0
5	25531		14.321	14.350	–	–	0	0
5	DD5		14.425		–	–	0	0
6	A25508	14.000	*		–	–	1	0
6	25508		14.185	14.500	–	–	0	0
6	DD6		14.341		–	–	0	0
7	A25441	14.000	*		–	–	1	0
7	25441		14.202	14.500	–	–	0	0
7	DD7		14.296		–	–	0	0
9	A25527	14.000	14.168		0.168056	–	0	1
9	25527		14.371	14.500	–	–	0	0
9	DD9		14.439		–	–	0	0
12	A25530	14.000	14.144		0.143923	–	1	0
12	25530		14.238	14.500	–	–	0	0
12	DD12		14.367		–	–	0	0

The results (see Table 5.8) after a few thousands nodes are $z := c^T + c^L = 1.334780$ with $c^T = 1.316780$ and an integrality gap of $\Delta = 8.5\%$. Note that this is again the same value, c^T, for the sum of lateness as in the previous example. But this time, there is no more than one lying passenger in a vehicle. Numerically, the inclusion of the load term into the objective function had the consequence that the lower bound, z^-, increased very slowly.

We also did further experiments, with a driving cost term included into the objective function (4.3.1) with $C^D_{n_1 n_2 v} = 0.001$. Our intention was to investigate whether this would destroy the intrinsic symmetry in the problem. The effect was, however, very small and the integrality gap even increased.

While these examples show that the MILP model can compute feasible solutions even in short time, the time to prove optimality is by far too large for implementing this approach in real world applications.

5.2.4   Column Enumeration Experiments

In this section we discuss the performance of our column enumeration approach (CEA) when applied to the online example described on page 129, and present one additional example.

Table 5.9 shows the results of the CEA for the test case TC9RC; the character R indicates that the restricted availability of the vehicles has been considered, C stands for CEA. The penalty parameter π_3 has been set to zero, *i.e.*, it might happen that a passengers spends more than 30 minutes in a vehicle. The objective function value, the sum of the lateness, is $z = 1.317333$, while the maximal lateness which occurred was 0.526 for vehicle 5 associated with the delivery of order A25364. It took only 14 seconds to compute this optimal solution involving $C_{6,3} = 140$ columns. Note that the tolerance parameters for the time availability of the vehicle have been set to 15 minutes (0.25 hours). Therefore, the tours for vehicle 2 and 3, which are actually only available till 14.20, are considered as feasible although the arrival time of vehicle 2 at the last delivery node is 14.203, which is just feasible.

Table 5.9: Optimal tours with restricted availability of the vehicles

v	s-node	d-node	tD	TD	tA	TT	delay	z	S	L	IC
1	SLOC80	A25528	14.128	0.015	14.143	13.766	0.377	0.377	0	1	0
1	A25528	25528	14.243	0.065	14.309	14.266	0.043	0.420	0	0	10
2	SLOC80	A25525	*	*	*	*	–	0.000	0	1	5
2	A25525	25525	14.128	0.075	14.203	14.500	–	0.000	0	0	10
3	SLOC54	A25532	*	*	*	*	–	0.000	0	1	5
3	A25532	25532	14.128	0.025	14.153	14.388	–	0.000	0	0	7
4	SLOC87	A25524	*	*	*	*	–	0.000	1	0	5
4	A25524	A25504	*	*	*	*	–	0.000	2	0	20

```
 4 A25504   25504    14.128  0.062 14.190 14.250    -     0.000 1 0 24
 4 25504    25524    14.240  0.079 14.319 14.260  0.059   0.059 0 0 17
 4 25524    A25533   14.419  0.074 14.500 14.500    -     0.059 0 1  0
 4 A25533   25533    14.550  0.075 14.625 15.000    -     0.059 0 0  8

 5 SLOC34   A25501      *       *     *      *       -     0.000 1 0  5
 5 A25501   A25531      *       *     *      *       -     0.000 2 0  5
 5 A25531   25501    14.128  0.061 14.189 14.250    -     0.000 1 0  9
 5 25501    25531    14.239  0.081 14.320 14.350    -     0.000 0 0 17
 5 25531    A25364   14.370  0.039 14.450 14.450    -     0.000 0 1  0
 5 A25364   25364    14.950  0.076 15.026 14.500  0.526   0.526 0 0 35

 6 SLOC-5   A25508      *       *     *      *       -     0.000 1 0  5
 6 A25508   A25534   14.128  0.088 14.250 14.250    -     0.000 1 1  0
 6 A25534   25508    14.350  0.051 14.401 14.500    -     0.000 0 1 22
 6 25508    25534    14.501  0.025 14.526 14.750    -     0.000 0 0 17

 7 SLOC-5   A25441      *       *     *      *       -     0.000 1 0  5
 7 A25441   25441    14.128  0.074 14.202 14.500    -     0.000 0 0 10

 9 SLOC55   A25527   14.128  0.040 14.168 14.000  0.168   0.168 0 1  0
 9 A25527   25527    14.318  0.053 14.371 14.500    -     0.168 0 0 13

12 SLOC73   A25530   14.128  0.016 14.144 14.000  0.144   0.144 1 0  0
12 A25530   25530    14.194  0.045 14.238 14.500    -     0.144 0 0  6
```

Note that an asterisk in the Table 5.9 indicates pre-assigned orders. For the pickup nodes of such pre-assigned orders the arrival times is not displayed.

To compare the CEA approach with the MILP approach, we applied it to the example TC9RC analyzed in Section 5.2.3.2, *i.e.*, we solved TC9RC under the assumption that the vehicles are available the whole day. In this case, we obtained $z = 1.317333$ with the tours shown in Table 5.10.

Table 5.10: Optimal tours with restriction on vehicle availability

v	s-node	d-node	tD	TD	tA	TT	delay	z	S	L	IC
1	SLOC80	A25528	14.128	0.015	14.143	13.766	0.377	0.377	0	1	0
1	A25528	25528	14.243	0.065	14.309	14.266	0.043	0.420	0	0	10
2	SLOC80	A25525	*	*	*	*	-	0.000	0	1	5
2	A25525	25525	14.128	0.075	14.203	14.500	-	0.000	0	0	10

3	SLOC54	A25532	*	*	*	*	–	0.000	0 1	5
3	A25532	25532	14.128	0.025 14.153	14.388	–	0.000	0 0	7
3	25532	A25534	14.203	0.041 14.250	14.250	–	0.000	0 1	0
3	A25534	25534	14.350	0.074 14.424	14.750	–	0.000	0 0	11
4	SLOC87	A25524	*	*	*	*	–	0.000	1 0	5
4	A25524	A25504	*	*	*	*	–	0.000	2 0	20
4	A25504	25504	14.128	0.062 14.190	14.250	–	0.000	1 0	24
4	25504	25524	14.240	0.079 14.319	14.260	0.059	0.059	0 0	17
5	SLOC34	A25501	*	*	*	*	–	0.000	1 0	5
5	A25501	A25531	*	*	*	*	–	0.000	2 0	5
5	A25531	25501	14.128	0.061 14.189	14.250	–	0.000	1 0	9
5	25501	25531	14.239	0.081 14.320	14.350	–	0.000	0 0	17
5	25531	A25364	14.370	0.039 14.450	14.450	–	0.000	0 1	0
5	A25364	25364	14.950	0.076 15.026	14.500	0.526	0.526	0 0	35
6	SLOC-5	A25508	*	*	*	*	–	0.000	1 0	5
6	A25508	25508	14.128	0.057 14.185	14.500	–	0.000	0 0	9
6	25508	A25533	14.285	0.025 14.500	14.500	–	0.000	0 1	0
6	A25533	25533	14.550	0.075 14.625	15.000	–	0.000	0 0	8
7	SLOC-5	A25441	*	*	*	*	–	0.000	1 0	5
7	A25441	25441	14.128	0.074 14.202	14.500	–	0.000	0 0	10
9	SLOC55	A25527	14.128	0.040 14.168	14.000	0.168	0.168	0 1	0
9	A25527	25527	14.318	0.053 14.371	14.500	–	0.168	0 0	13
12	SLOC73	A25530	14.128	0.016 14.144	14.000	0.144	0.144	1 0	0
12	A25530	25530	14.194	0.045 14.238	14.500	–	0.144	0 0	6

Note that this value is the same as in the more restrictive case described above. The MILP model gave a slightly different objective function value $z = 1.31678$. The difference of 5.53×10^{-4} is caused by numerical rounding.

Table 5.9 shows that only 9 vehicles are needed. Using the same set of columns it is easy to investigate how the objective function value increases if less vehicles are allowed or available. As there are already 6 vehicles with pre-assigned orders, only the cases $6 \leq N^V \leq 9$ leading to the results

N^V	9	8	7	6
z	1.317333	1.337800	1.383455	1.459955

are of interest. In this example, the results were obtained within seconds. Allowing a total increase of about 11% for the sum of the delays, 3 vehicles could be saved completely.

Let us now solve another online example with $N_A = 13$ pre-assigned and $N_F = 13$ free orders to be assigned to 10 vehicles. This set of orders involves one, A64439, requiring a wheelchair passenger needs to be transported, and another one, A64453, which just needs to carry some goods not requiring any capacity. In this case study the penalty parameter controlling that a passenger does not spend more than $\Delta T = 30$ minutes in a vehicle, has been set to a high value of 200.

The solution table below indicates pre-assigned orders by an asterisk in some lines; the same line shows in the last column how many minutes the passengers is already in the car when the clock is set to the current time of $T_* = 11.004$. Vehicle 4 has three pre-assigned orders. With 13 orders to be distributed to 10 vehicles, we set the number, m, of orders per subset to $m = 5$, *i.e.*, those vehicles with no pre-assigned orders can pickup at most 5 orders. The columns for vehicles containing pre-assigned orders are allowed to have a total of $m + 1$, *i.e.*, 6 orders.

Note that wheelchair orders are treated as loading two sitting passengers. Due to the strong penalization of having passengers spending more than 30 minutes in a car, the orders close to that limits are always delivered before other orders. It turns out in the optimal solution that all vehicles have about the same load (number of orders).

The total number of columns generated was $C_{13,6} = 12030$. It took 20 seconds to generate all of them, but 44 minutes to evaluate them. The set partitioning problem was solved within 3 seconds giving the optimal solution with an objective function value of $z = 0.382422$. None of the 10 vehicles had a total of more than 3 orders in the optimal solution, except for vehicle 4 which had already 3 pre-assigned orders and got one additional order assigned. Therefore, we did a comparative run based on $C_{13,3} = 3779$ columns. Within 10 seconds an optimal solution with the same objective function value but a different assignment of the vehicles was obtained.

As the computational time of 44 minutes is too much time for the online dispatcher, the numerical experiments based on $C_{13,5}$ and $C_{13,3}$ triggered those improvements described in Section 4.3.3.6 using the cut-off value, Z_C, in the routing-scheduling tree. Using this approach the problem is solved within 4 minutes giving the same solution with objective function value $z = 0.382422$. The detailed solution is shown in Table 5.11.

Table 5.11: Online optimization with 13 pre-assigned and 13 free orders

v	s-node	d-node	tD	TD	tA	TT	delay	z	S	L	IC
1	SLOC60	A64441	*	*	*	*	–	0.000	0	1	13
1	A64441	A64439	11.004	0.058	11.062	11.000	0.062	0.062	2	1	0
1	A64439	64441	11.178	0.059	11.237	11.300	–	0.062	2	0	27
1	64441	64439	11.370	0.052	11.422	11.500	–	0.062	0	0	22
2	SLOC70	A64432	*	*	*	*	–	0.000	0	1	16
2	A64432	A64447	11.004	0.060	11.064	11.000	0.064	0.064	1	1	0
2	A64447	64432	11.114	0.047	11.161	11.183	–	0.064	1	0	25
2	64432	64447	11.244	0.066	11.310	11.500	–	0.064	0	0	15
3	SLOC57	A64449	*	*	*	*	–	0.000	1	0	10
3	A64449	A64456	11.004	0.074	11.250	11.250	–	0.000	1	1	0
3	A64456	64449	11.283	0.000	11.283	11.400	–	0.000	0	1	27
3	64449	64456	11.383	0.003	11.387	11.750	–	0.000	0	0	9
4	SLOC57	A64435	*	*	*	*	–	0.000	1	0	16
4	A64435	A64433	*	*	*	*	–	0.000	2	0	20
4	A64433	A64438	*	*	*	*	–	0.000	2	1	16
4	A64438	64435	11.004	0.043	11.047	11.250	–	0.000	1	1	18
4	64435	64433	11.064	0.088	11.151	11.200	–	0.000	0	1	29
4	64433	64438	11.151	0.067	11.219	11.267	–	0.000	0	0	29
4	64438	A64337	11.369	0.000	11.500	11.500	–	0.000	0	1	0
4	A64337	64337	11.633	0.040	11.674	12.000	–	0.000	0	0	11
5	SLOC80	A64430	*	*	*	*	–	0.000	1	0	20
5	A64430	A64451	11.004	0.000	11.004	11.000	0.004	0.004	1	1	0
5	A64451	64430	11.037	0.055	11.092	11.250	–	0.004	0	1	26
5	64430	A64454	11.092	0.017	11.250	11.250	–	0.004	1	1	0
5	A64454	64451	11.300	0.060	11.360	11.500	–	0.004	1	0	22
5	64451	64454	11.443	0.035	11.478	11.750	–	0.004	0	0	14
6	SLOC23	A64322	11.004	0.032	11.036	11.000	0.036	0.036	1	0	0
6	A64322	A64453	11.036	0.011	11.047	11.000	0.047	0.083	1	0	0
6	A64453	A64320	11.047	0.021	11.068	11.000	0.068	0.150	2	0	0
6	A64320	64320	11.134	0.044	11.179	11.500	–	0.150	1	0	7
6	64320	64322	11.245	0.000	11.245	11.500	–	0.150	0	0	13
6	64322	64453	11.279	0.062	11.341	11.500	–	0.150	0	0	18
7	SLOC47	A64436	*	*	*	*	–	0.000	0	1	10
7	A64436	A64440	11.004	0.070	11.074	11.000	0.074	0.074	1	1	0
7	A64440	64436	11.157	0.070	11.227	11.250	–	0.074	1	0	24
7	64436	64440	11.344	0.097	11.441	11.500	–	0.074	0	0	23

8	SLOC71	A64429	*	*	*	*	–	0.000	0 1 16
8	A64429	64429	11.004	0.058	11.062	11.066	–	0.000	0 0 19
8	64429	A64455	11.162	0.043	11.250	11.250	–	0.000	0 1 0
8	A64455	64455	11.400	0.045	11.445	11.750	–	0.000	0 0 12
9	SLOC85	A64356	*	*	*	*	–	0.000	0 1 10
9	A64356	A64450	*	*	*	*	–	0.000	0 2 26
9	A64450	64450	11.004	0.045	11.049	11.500	–	0.000	0 1 28
9	64450	64356	11.240	0.039	11.279	11.250	0.029	0.029	0 0 27
9	64356	A64431	11.363	0.045	11.500	11.500	–	0.029	1 0 0
9	A64431	64431	11.550	0.057	11.607	12.000	–	0.029	0 0 7
10	SLOC34	A64434	*	*	*	*	–	0.000	0 1 15
10	A64434	A64444	*	*	*	*	–	0.000	1 1 10
10	A64444	64434	11.004	0.020	11.023	11.500	–	0.000	1 0 16
10	64434	A64443	11.173	0.066	11.250	11.250	–	0.000	1 1 0
10	A64443	64444	11.317	0.016	11.333	11.350	–	0.000	0 1 30
10	64444	64443	11.416	0.045	11.460	11.750	–	0.000	0 0 13

5.2.5 Construction and Improvement Heuristics

Here we present the results of several experiments for the implemented heuristics described in Section 4.4. For the numerical experiments we use the real data recorded on September 1, 18 and 19, 2001 (for the input data see Appendix B.B.2.2.1). We tested our heuristics in an offline situation (as we know all orders in the morning in advance) and in an online situation (the algorithm is applied subsequently, when 10 new orders have been collected). In our program, online and offline cases are controlled by parameter IOPTMD (see Appendix for the possible values of all control parameters). The quality of the tours is characterized by the quantities listed on page 69:

1 the total lateness, $z = \sum_{v=1}^{N^V} S_v^D$;

2 the sum of maximal lateness, S_D;

3 the average sum of lateness, $\sum_{v=1}^{N^V} z_v / N^V$;

4 the average maximum lateness, $\bar{S}_D$;

5 the maximal lateness, D_*.

5.2.5.1 Offline Version

We test and compare all combinations of the heuristic steps CONH1, CONH2 with IT1, IT2 and distribution of non-assigned orders using DB. We do not use DU because in most offline cases there are many orders that are originally not assigned to any vehicle. If these unassigned order would be distributed uniformally over the vehicles, too many infeasibilities would be produced. When applying the combination IT1 and IT2 heuristics, we test the following acceptance criteria:

1 a new tour for a vehicle v is accepted, if:

$$((S_v^D < S_{0v}^D)\&(D_v < D_{0v}))\text{ or}$$
$$((S_v^D < 0.8 * S_{0v}^D)\&(D_v < 1.05 * D_{0v}))\text{ or}$$
$$((S_v^D < 1.05 * S_{0v}^D)\&(D_v < 0.8 * D_{0v}));$$

2 a new tour is accepted unconditionally.

Thus, we have eight possible combinations of how to combine the steps CONH1, CONH2 with the steps IT1, IT2:

I:	CONH1 and IT1
II:	CONH1 and IT1+IT2 with the first acceptance criteria
III:	CONH1 and IT1+IT2 with the second acceptance criteria
IV:	CONH1 and IT2
V:	CONH2 and IT1
VI:	CONH2 and IT1+IT2 with the first acceptance criteria
VII:	CONH2 and IT1+IT2 with the second acceptance criteria
VIII:	CONH2 and IT2

The results of various experiments are summarized in Tables 5.12, 5.13 and 5.14 using the real data for 1, 18 and 19 September, 2001 (see Appendix B.B.2.2.1).

Table 5.12: Comparison of the algorithms for September, 1

Char./Alg.	I	II	III	IV	V	VI	VII	VIII
1	8.27	8.80	11.29	16.65	8.27	8.80	11.29	16.65
2	1.90	2.10	2.08	2.75	1.90	2.10	2.10	2.75
3	0.69	0.73	0.94	1.39	0.69	0.73	0.73	1.39
4	0.16	0.17	0.17	0.23	0.16	0.17	0.17	0.23
5	0.65	0.55	0.55	0.55	0.65	0.55	0.55	0.55
CPU, sec.	6.60	14.31	10.30	5.12	6.79	14.41	10.29	5.16

Table 5.13: Comparison of the algorithms for September, 18

Char./Alg.	I	II	III	IV	V	VI	VII	VIII
1	19.36	20.54	17.21	23.63	19.36	20.54	17.79	23.63
2	3.47	2.97	3.10	3.09	3.47	2.97	2.78	3.09
3	1.76	1.87	1.56	2.15	1.76	1.87	1.62	2.15
4	0.31	0.27	0.28	0.28	0.32	0.27	0.25	0.28
5	0.49	0.53	0.38	0.58	0.49	0.53	0.54	0.58
CPU, sec.	21.99	43.85	92.82	25.22	22.31	44.38	39.17	25.66

Table 5.14: Comparison of the algorithms for September, 19

Char./Alg.	I	II	III	IV	V	VI	VII	VIII
1	21.18	24.17	27.18	40.22	25.44	25.80	35.44	35.92
2	3.08	2.82	3.44	3.52	3.35	2.88	2.76	2.92
3	1.93	2.20	2.47	3.66	2.31	2.34	3.22	3.27
4	0.28	0.26	0.31	0.32	0.30	0.26	0.25	0.26
5	0.66	0.53	0.74	0.79	0.67	0.65	0.90	0.72
CPU, sec.	25.07	59.62	357.62	24.19	22.68	34.18	75.18	23.76

The results in the tables above lead us to the conclusion that there is no best heuristic which outperforms all other for all data instances. What in average performed best was a combination of IT1 and IT2 with either CONH1 or CONH2.

5.2.5.2 Online version

The online version of the construction and improvement heuristic is only slightly different from the offline version. Instead of the depot, any position is used as the original location of a vehicle. Furthermore, there is no need to return to the depot. All information about the original locations of the vehicles, their availability and the current time is located in a file displayed in Fig. B.1.2.

In Table 5.15 we show the results of our heuristic approaches for the test case TC9RH; the character R indicates the restricted availability of the vehicles has been considered, H stands for heuristics. This example case is the same as TC9RC considered in Section 5.2.4. Using the heuristics, the objective function value, *i.e.*, the total lateness is $z = 1.537877$, while the maximal lateness which occurred was 0.528155 for vehicle 6 (again the delivery of order A25364). Note that the objective function value differs approximately by 17%, *i.e.*, 0.22 from that of the optimal solution derived in Section 5.2.4.

Table 5.15: Results of experiments for the test example TC9RH

```
v s-node  d-node     tD     TD     tA     TT    delay    z     S L IC
----------------------------------------------------------------------
1 SLOC80  A25534  14.175  0.075 14.250 14.250    -    0.000 0 1  0
1 A25534  25534   14.350  0.074 14.424 14.750    -    0.000 0 0 11
1 25534   A25533  14.474  0.000 14.500 14.500    -    0.000 0 1  0
1 A25533  25533   14.550  0.075 14.625 15.000    -    0.000 0 0  8

2 SLOC80  A25525    *       *      *      *       -    0.000 0 1  5
2 A25525  25525   14.128  0.075 14.203 14.500    -    0.000 0 0 10
```

3	SLOC54	A25532	*	*	*	*	–	0.000	0 1	5
3	A25532	25532	14.128	0.025	14.153	14.388	–	0.000	0 0	7
4	SLOC87	A25504	*	*	*	*	–	0.000	1 0	20
4	A25504	A25524	*	*	*	*	–	0.000	2 0	5
4	A25524	25504	14.128	0.002	14.130	14.250	–	0.000	1 0	21
4	25504	25524	14.180	0.079	14.259	14.260	–	0.000	0 0	13
5	SLOC34	A25501	*	*	*	*	–	0.000	1 0	5
5	A25501	A25531	*	*	*	*	–	0.000	2 0	5
5	A25531	A25528	14.128	0.000	14.128	13.766	0.362	0.362	2 1	0
5	A25528	25501	14.228	0.061	14.289	14.250	0.039	0.401	1 1	15
5	25501	25531	14.339	0.081	14.420	14.350	0.070	0.471	0 1	23
5	25531	25528	14.470	0.000	14.470	14.266	0.204	0.675	0 0	21
6	SLOC-5	A25508	*	*	*	*	–	0.000	1 0	5
6	A25508	A25530	14.128	0.036	14.164	14.000	0.164	0.164	2 0	0
6	A25530	25508	14.214	0.027	14.241	14.500	–	0.164	1 0	12
6	25508	25530	14.341	0.018	14.359	14.500	–	0.164	0 0	12
6	25530	A25364	14.409	0.043	14.452	14.450	0.002	0.167	0 1	0
6	A25364	25364	14.952	0.076	15.028	14.500	0.528	0.695	0 0	35
7	SLOC-5	A25441	*	*	*	*	–	0.000	1 0	5
7	A25441	25441	14.128	0.074	14.202	14.500	–	0.000	0 0	10
9	SLOC55	A25527	14.128	0.040	14.168	14.000	0.168	0.168	0 1	0
9	A25527	25527	14.318	0.053	14.371	14.500	–	0.168	0 0	13

If it is assumed that the vehicles are available the whole day (test case
TC9H), the heuristic produces a set of tours with $z = 1.352643$ and the
following tours (see Table 5.16).

Table 5.16: Results of the test case TC9H

v	s–node	d–node	tD	TD	tA	TT	delay	z	S L	IC
1	SLOC80	A25528	14.128	0.015	14.143	13.766	0.377	0.377	0 1	0
1	A25528	25528	14.243	0.065	14.309	14.266	0.043	0.420	0 0	10
1	25528	A25364	14.459	0.039	14.497	14.450	0.047	0.467	0 1	0
1	A25364	25364	14.997	0.076	15.073	14.500	0.573	1.041	0 0	35
2	SLOC80	A25525	*	*	*	*	–	0.000	0 1	5
2	A25525	25525	14.128	0.075	14.203	14.500	–	0.000	0 0	10
3	SLOC54	A25532	*	*	*	*	–	0.000	0 1	5
3	A25532	25532	14.128	0.025	14.153	14.388	–	0.000	0 0	7

4	SLOC87	A25504	*	*	*	*	–	0.000	1	0	20
4	A25504	A25524	*	*	*	*	–	0.000	2	0	5
4	A25524	25504	14.128	0.002	14.130	14.250	–	0.000	1	0	21
4	25504	25524	14.180	0.079	14.259	14.260	–	0.000	0	0	13
5	SLOC34	A25501	*	*	*	*	–	0.000	1	0	5
5	A25501	A25531	*	*	*	*	–	0.000	2	0	5
5	A25531	25501	14.128	0.061	14.189	14.250	–	0.000	1	0	9
5	25501	25531	14.239	0.081	14.320	14.350	–	0.000	0	0	17
6	SLOC-5	A25508	*	*	*	*	–	0.000	1	0	5
6	A25508	25508	14.128	0.057	14.185	14.500	–	0.000	0	0	9
7	SLOC-5	A25441	*	*	*	*	–	0.000	1	0	5
7	A25441	25441	14.128	0.074	14.202	14.500	–	0.000	0	0	10
9	SLOC55	A25527	14.128	0.040	14.168	14.000	0.168	0.168	0	1	0
9	A25527	25527	14.318	0.053	14.371	14.500	–	0.168	0	0	13

In this case, the difference to the optimal solution with $z = 1.3173$ is again only very small. The large difference obtained for the test case TC9H indicates that the use of penalty strategies is not ideal for the inter-tour heuristic; it seems to be difficult to reassign time-critical orders to a different vehicle.

5.2.6   Simulated Annealing Examples

Simulated annealing when applied to a problem requires to specify the following tuning parameters:

- the initial tuning temperature, T_0,

- how many iterations, N_T, are performed at each temperature,

- how much the temperature is decreased at each step as cooling proceeds described by either ΔT or δ, and

- the frequency ratio, ρ, of swaps and shifts.

In a series of initial numerical experiments we derived some appropriate values for the tuning parameters ρ and δ involved in the simulated annealing method described in Section 4.4.3. Usually, $\rho = 0.0125$ is a good value to select *shifts* or *swaps*, while $\delta = 0.9$ should be chosen to connect subsequently chosen temperatures. The initial temperature, T_0, is set to $T_0 = 0.1z_0$, where z_0 is the objective function value of the tours initially provided. The temperature is decreased for at most $N_T = 100$ times.

In this section we apply SA to three examples. For these examples, we tested the following combinations:

A: CIH only; no SA (IFSA1=IFSA2= 0)
B: SA to each individual tour after applying CIH; (IFSA1= 0, IFSA2= 1)
C: after B, the optimal tours are used as input to the CIH, followed by a final SA step.

For the three September data sets (scenarios S_2, S_f, and S_g) we got the following results:

	Sep 1, 2001 (S_2)			Sep 18, 2001 (S_f)			Sep 19, 2001 (S_g)		
τ	A	B	C	A	B	C	A	B	C
1	8.80	5.95	5.76	17.21	16.44	14.50	27.18	29.31	29.93
2	2.10	1.53	1.68	3.10	2.32	2.50	3.44	3.79	3.07
3	0.73	0.50	0.48	1.56	1.49	1.32	2.47	2.66	2.72
4	0.17	0.13	0.14	0.28	0.21	0.23	0.31	0.34	0.28
5	0.55	0.29	0.44	0.38	0.36	0.36	0.74	0.74	1.00
	48	872	1310	92	1363	2453	358	2940	2940

where in each row the quantities introduced at the beginning of Section 5.2.5 are given, and τ is the CPU time in seconds. The first heuristic (A) corresponds to combination II (for Sep 1) and combination III (Sep 18 and 19) described in the previous subsection. The last combination of all heuristics (C) which might be considered less good because of the strongly increased value of the maximum lateness. Thus, the overall conclusion is that SA, applied to single-tour optimization and used at the right places, can improve the quality of offline solutions covering a whole day.

5.2.7   Summary of the Numerical Experiments

In this section we have applied various algorithms to offline and online problems. While the offline problems with several hundred orders can only be solved with heuristics and metaheuristics, we are able to find optimal solutions for the online cases, relevant to our real world problem, in short time. The taylor-suited heuristics produces significantly better results than the simulated annealing method for the overall problem, however, SA can improve existing single-vehicle tours or can lead to better solutions in combination with the heuristic. The MILP model produces optimal solutions only in small and special cases with less than 5 or 10 orders. In some cases, we also have solved intra-tour optimization problems with up to 16, sometimes even 25, orders to optimality using the MILP approach. Although the MILP model can solve problems with a smaller number of orders to optimality, it depends too strongly on the data structure and fine tuning and thus is not appropriate to be implemented into a decision support tool which should deliver a good quality solutions in all situations. Optimal solutions of online problems including up to 20 or 30 orders to be assigned to a fleet of 10 vehicles are obtained by a column enumeration approach. The exact solutions are further used to evaluate the quality of our heuristic intra-tour algorithms. This shows that the heuristic methods in many cases produce optimal or near-optimal solutions.

Table 5.17 shows the range of the number of orders when to apply which method to solve the intra-tour optimization problem.

Table 5.17: The performance of the solution approaches for solving the intra-tour problem as a function of the number of orders.

	1-5	5-8	8-15	15-40
MILP	√	(√)	(√)	
CIH	√	√	*	(√)
B&B	*	*		
SA	√	√	√	√
CIH+SA	√	√	√	*

In Table 5.17, the preferred method for each range of number of orders is indicated by "*". The symbol "√" means, that for this range the method works reasonably but may be not the most efficient one; "(√)" means that this may be possible but is not really recommended due to its low efficiency.

Table 5.18 shows the range for the ratio, N_F/N^V, free orders, N_F, to be assigned over the number, N^V, of available vehicles. It gives the optimal range for solving the overall problem including the assignment and the sequencing problem.

Table 5.18: The performance of the solution approaches for solving the complete VRPPDTW (assignment, sequencing, and scheduling) as a function of the ratio: free orders over available vehicles.

	1-5	5-8	8-15	15-40
MILP	√	(√)	(√)	
CEA+B&B	*	*		
CEA+CIH			(√)	
CEA+CIH+SA				(√)
CIH	√	√	*	(√)
SA	√	√	√	√
CIH+SA	√	√	√	*

Note that the range $N_F/N^V \leq 8$ covers most of the online scenarios. In rush-hours about 20 to 30 new orders arrive per hour. Usually, rescheduling is done after each 10 minutes. With a fleet of about 10 vehicles, the ratio $N_F/N^V \leq 8$ is easily met. The range 15-40 corresponds rather to the offline scenarios.

The sequence of online solutions obtained during a day can be compared to the approximate solutions derived from the input data of all orders assumed to be known in advance. A key quantity to inspect is the average maximum delay. The numerical experiments covering a whole day in Section 5.2.5 and 5.2.6 show that the average maximum delay is of the order of 0.15 hours, *i.e.*, nine minutes. This knowledge and the usage of our online strategy can

- reduce the waiting times for passenger's pickup,

- decrease delayed pickup and delivery,

- reduce the time a passenger spends in a car, and

- lead to a better load balance of all vehicles.

Some of the algorithms, especially the column enumeration approach discussed here, can be embedded into a dispatcher support system for a hospital. The construction and improvement heuristics have already been implemented as a decision support tool used for simulation purposes. The heuristics alone lead already to significant reductions of the passenger waiting time by at least 12% (see Fig. 5.2.5) and vehicle driving times (see Fig. 5.2.6). More details are given in [22, Section 2.3].

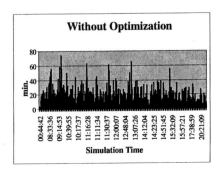 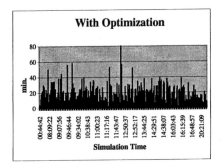

Figure 5.2.5. Waiting times during the course of a day

5.3. Summary

The quality of the online tours produced by our heuristics has been compared to the exact optimal solution where available. The comparison

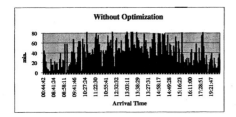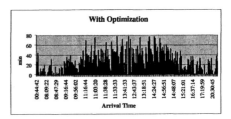

Figure 5.2.6. Driving times during the course of a day

shows that for small and medium size scenarios the heuristic solutions are often close to the optimal solution. However, if soft constraints, (*e.g.*, the vehicle availability constraints) are modeled by penalty functions the solution quality decreases. For larger offline scenarios the solution produced by the heuristic method are improved by simulated annealing. Thus, there is some room to improve the heuristic methods.

One possible approach to solve the problem faster is to use a combination of the MILP model and heuristic approaches. Because solving the MILP problem or proving optimality of its solutions may take too long, it could be reasonable to treat the assignment part, *i.e.*, distributing orders over all vehicles, using some heuristic (for instance, CONH1 or CONH2). The heuristic step gives the assignment of the orders to the vehicles and the upper bounds on the sum of lateness. This upper bound is exploited by the MILP model or the branch-and-bound approach to compute the optimal solution of the intra-tour problem.

Alternatively, full column generation techniques involving a branch-and-price algorithm can be applied to solve larger scenarios. The current version is based on column *enumeration* generating all possible columns in advance - this prevents us from solving large problems to optimality. For the offline case to check the quality of solutions produced by our heuristics it can be useful to apply a column generation approach, while for solving subproblems of smaller size - to use our suggested branch-and-bound approach.

Finally, we want to comment on the quality of the online solutions compared to the offline solutions. Instead of a competitive analysis we have estimated the quality of our online algorithms by comparing the characteristic properties of the online and offline solutions. The most relevant quantity is the average delay per order. The values of the average delay of orders are similar in the offline and online solutions. More cannot be expected.

Chapter 6

SUMMARY

In this book we have analyzed and developed online algorithms involving exact optimization and heuristic techniques to support decision making in inventory and transportation logistics. We have applied those methods to two real life problems: a special carousel based high-speed storage system - *Rotastore* and a hospital transportation problem.

In Chapter 2 we showed that some types of logistics problem are NP-hard and not easy to solve to optimality in offline situations. We considered a polynomial case and developed an exact algorithm to solve a certain type of the Batch PreSorting Problem (BPSP) in offline situations. To estimate the quality of our algorithms developed for online cases we use competitive analysis showing that the proposed online algorithm is $\frac{3}{2}$-competitive. To check how the knowledge of future requests can improve the performance of the algorithm, we constructed new online algorithms with lookahead. Some of them are able to improve the online solutions. Moreover, we investigated the origin of the problem complexity and proved that the problem without the capacity constraint on an additional storage has an integer polyhedron.

The vehicle routing problem we solved for the health sector is a typical online optimization problem (Chapter 4). It was demonstrated that reasonable solutions for the offline case covering a whole day with a few hundred orders can be constructed with a heuristic approach, as well as by simulated annealing. For small cases our MILP model and exact optimization works as well but it is impractical in terms of running times.

Optimal solutions for small instances (up to ten additional orders per vehicle) are computed by a column enumeration approach (CEA) based on decomposing the problem into a master problem (a set partitioning problem controlling the assignment of orders to vehicles) and a set of subproblems (intra-tour problem, routing and scheduling). To solve the routing and scheduling problem exactly we have developed a branch-and-bound method which prunes nodes if they are value-dominated by previous found solutions or if they are infeasible *w.r.t.* to the capacity constraints. This approach works fast for up to 10 orders per vehicle. This branch-and-bound method is suitable to solve any kind of sequencing-scheduling problem involving cumulative objective functions and constraints, which can be evaluated sequentially. The CEA we have developed to solve the hospital transportation problem is of general nature. Thus, it can be embedded into other decision-support systems involving assigning, sequencing and scheduling, if the size of the problem and the structure of the sequencing and scheduling problem problem are appropriate.

Further work in this direction might focus on solving exactly larger instances of the subproblems with larger numbers of orders per vehicle. The comparison of the results obtained by the heuristic methods with those for which we were able to compute the optimal solution shows that the heuristic in many cases produces optimal or near-optimal solutions.

What can a logistics manager responsible for an inventory storage systems and the developer of decision support systems for a hospital campus learn from this book? They can learn that their problems can be solved using online optimization techniques. The methods and algorithms we have developed can be adapted to solve similar logistic problems. And there is good reason to believe that they can improve the quality and reliability of the decisions significantly.

GLOSSARY

Arc: An object within a graph. Arcs, sometimes also called edges, usually represent roads, pipelines, or similar paths along which some material can flow. Often arcs have a capacity. Arcs connect the nodes in a graph.

Basic variables: Those variables in optimization problems whose values, in non-degenerate cases, are away from their bounds and are uniquely determined from a system of equations.

Basis (Basic feasible solution): In an LP problem with constraints $Ax = b$ and $x \geq 0$ the set of m linearly independent columns of the $m \times n$ system matrix A of an LP problem with m constraints and n variables forming a regular matrix $\mathcal{B}$. The vector $x_B = \mathcal{B}^{-1}b$ is called a basic solution. x_B is called a basic feasible solution if $x_B \geq 0$.

Bound: Bounds on variables are special constraints. A bound involves only one variable and a constant which fixes the variable to that value, or serves as a lower or upper limit.

Branch & Bound: An implicit enumeration algorithm for solving combinatorial problems. A general Branch & Bound algorithm for MILP problems operates by solving an LP relaxation of the original problem and then performing a systematic search for an optimal solution among sub-problems formed by branching on a variable which is not currently at an integer value to form a sub-problem, resolving the sub-problems in a similar manner.

Branch & Cut: An algorithm for solving mixed integer linear programming problems which operates by solving a linear program which is a relaxation of the original problem and then performing a systematic search for an optimal solution by adjoining to the relaxation a series of valid constraints (cuts) which must be satisfied by the integer aspects of the problem to the relaxation, or to sub-problems generated from the relaxation, and resolving the problem or sub-problem in a similar manner.

Branch & Price: An algorithm for solving large structured mixed integer linear programming problems. This algorithms is usually part of a column generation algorithms.

Column enumeration: An algorithm for solving large mixed integer linear programming problems which operates by solving a master problem and a set of subproblems. Unlike column generation, in this case all columns are generated and evaluated a priori. The master problem is often a set partitioning problem. Column enumeration is a special case of column generation.

Column generation: An algorithm for solving large linear or mixed integer linear programming problems which operates by solving a master problem and then generating new columns (variables or more general objects) by solving a subproblem. If the subproblem contains integer variables, a Branch & Price algorithm is used in addition.

Constraint: A relationship that limits implicitly or explicitly the values of the variables in a model. Usually, constraints are formulated as inequalities or equations representing conditions imposed on a problem, but other types of relations exist, *e.g.*, set membership relations.

Feasible point (feasible problem):   A point (or vector) to an optimization problem that satisfies all the constraints of the problem. (A problem for which at least one feasible point exists.)

Graph: A mathematical object consisting of nodes and arcs, useful in describing network flow problems. The structure and properties of graphs are analysed in graph theory, a mathematical discipline.

Heuristic solution: A feasible point of an optimization problem which is not necessarily optimal and has been found by a constructive technique which could not guarantee the optimality of the solution.

Infeasible problem: A problem for which no feasible point exists.

Node: An object within a graph. Nodes usually represent plants, depots, or a point in a network. Nodes can be connected by arcs.

$\mathcal{NP}$ completeness: Characterization of how difficult it is to solve a certain class of optimization problems. The computational requirements increase exponentially with some measure of the problem size.

Objective (objective function):   An expression in an optimization problem that has to be maximized or minimized.

Online Optimization: A sub-discipline of optimization applied to real life decision problems where decision should be made online based on partial, insufficient information or without any knowledge of future.

Offline Optimization: A technical term used in opposition to online optimization. It is assumed that all information about the future are given completely and deterministically.

Optimization: The process of finding the best solution (according to some criterion technically called objective function) of an optimization problem subject to constraints.

Optimum (optimal solution):   A feasible point of an optimization problem that cannot be improved on, in terms of the objective function, without violating the constraints of the problem.

Relaxation: An optimization problem created from another where some of the constraints have been removed or weakened.

Simplex algorithm: Algorithm for solving LP problems that investigates vertices of polyhedra.

Unimodularity: A property of a matrix. A squared matrix is called unimodular if its determinant is $+1$. An LP matrix is called unimodular if all its sub-matrices have determinants with value $+1$, 0 or -1. If an LP matrix is unimodular, and the right-side constraint vector has only integer entries, then all basic feasible solutions to the LP take integer values.

Variable: An algebraic symbol used to represent a decision or other varying quantity. Variables are also called "unknowns" or "columns".

Appendix A
Rotastore

A.1.   Tabular Results for Different Models

The first set of tables shows the results for BPSP$_1$ for test example M60-1 with $N^S = 2, ..., 5$.

Table: BPSP$_1$ for test example M60-1 with $N^S = 2$.

Cycle	NS	CPU,sec.	NOC	opt NOC	Optimality gap
1	2	2.453	45	44	2.27%
2	2	2.343	75	68	10.29%
3	2	2.383	93	79	17.72%
4	2	2.313	107	86	24.42%
5	2	2.153	124	88	40.91%
6	2	2.163	140	92	52.17%
7	2	2.173	154	98	57.14%
8	2	2.153	175	103	69.90%
9	2	2.203	187	112	66.96%
10	2	2.233	199	120	65.83%
11	2	2.403	210	128	64.06%
12	2	2.193	225	134	67.91%
13	2	2.193	242	142	70.42%
14	2	2.223	254	152	67.11%
15	2	2.223	262	161	62.73%
total:		33.805	262	161	av: 49.32%

Table: BPSP$_1$ for test example M60-1 with $N^S = 3$.

Cycle	NS	CPU,sec.	NOC	opt NOC	Optimality gap
1	3	2.274	44	44	0.00%
2	3	2.274	73	68	7.35%
3	3	2.153	91	79	15.19%
4	3	2.143	106	85	24.71%
5	3	2.163	123	87	41.38%
6	3	2.133	139	91	52.75%

7	3	2.133	158	95	66.32%
8	3	2.163	181	98	84.69%
9	3	2.193	194	105	84.76%
10	3	2.183	209	114	83.33%
11	3	2.323	220	125	76.00%
12	3	2.173	235	134	75.37%
13	3	2.133	255	140	82.14%
14	3	2.163	265	149	77.85%
15	3	2.173	272	156	74.36%

total:		32.777	272	156	av: 56.41%

Table: $BPSP_1$ for test example M60-1 with $N^S = 4$.

Cycle	NS	CPU,sec.	NOC	opt NOC	Optimality gap
1	4	2.313	44	44	0.00%
2	4	2.233	75	68	10.29%
3	4	2.153	96	79	21.52%
4	4	2.153	112	85	31.76%
5	4	2.154	127	87	45.98%
6	4	2.153	142	91	56.04%
7	4	2.173	156	95	64.21%
8	4	2.153	177	97	82.47%
9	4	2.163	190	102	86.27%
10	4	2.143	205	111	84.68%
11	4	2.153	215	122	76.23%
12	4	2.133	230	132	74.24%
13	4	2.173	248	139	78.42%
14	4	2.143	260	149	74.50%
15	4	2.173	266	155	71.61%

total:		32.566	266	155	av: 57.22%

Table: $BPSP_1$ for test example M60-1 with $N^S = 5$.

Cycle	NS	CPU,sec.	NOC	opt NOC	Optimality gap
1	5	2.364	44	44	0.00%
2	5	2.153	74	68	8.82%
3	5	2.383	94	79	18.99%
4	5	2.353	110	85	29.41%
5	5	2.123	126	87	44.83%
6	5	2.303	146	91	60.44%
7	5	2.113	158	96	64.58%
8	5	2.123	176	98	79.59%
9	5	2.313	192	103	86.41%
10	5	2.393	203	111	82.88%
11	5	2.333	218	122	78.69%
12	5	2.303	227	132	71.97%
13	5	2.193	245	138	77.54%
14	5	2.323	257	147	74.83%
15	5	2.143	269	155	73.55%

total:		33.916	269	155	av: 56.84%

The next set of tables shows the results for $BPSP_2$ for test example M60-1 with $N^S = 2, ..., 5$.

Table: $BPSP_2$ for test example M60-1 with $N^S = 2$.

Cycle	NS	CPU,sec.	NOC	opt NOC	Optimality gap
1	2	10.82	44	44	0.00%
2	2	7.03	68	68	0.00%
3	2	6.75	79	79	0.00%
4	2	6.81	85	86	-1.16%
5	2	6.75	87	88	-1.14%
6	2	7.20	99	92	7.61%
7	2	7.14	109	98	11.22%
8	2	6.82	117	103	13.59%
9	2	6.81	129	112	15.18%
10	2	6.70	135	120	12.50%
11	2	6.71	142	128	10.94%
12	2	6.98	148	134	10.45%
13	2	6.86	161	142	13.38%
14	2	6.81	170	152	11.84%
15	2	15.60	181	161	12.42%
total:		115.79	181	161 av:	7.79%

Table: $BPSP_2$ for test example M60-1 with $N^S = 3$.

Cycle	NS	CPU,sec.	NOC	opt NOC	Optimality gap
1	3	8.13	44	44	0.00%
2	3	7.36	68	68	0.00%
3	3	6.64	79	79	0.00%
4	3	6.97	85	85	0.00%
5	3	6.31	87	87	0.00%
6	3	6.75	94	91	3.30%
7	3	6.97	103	95	8.42%
8	3	6.75	110	98	12.24%
9	3	7.41	124	105	18.10%
10	3	7.58	128	114	12.28%
11	3	8.51	139	125	11.20%
12	3	7.96	147	134	9.70%
13	3	7.97	154	140	10.00%
14	3	9.28	161	149	8.05%
15	3	10.66	171	156	9.62%
total:		115.25	171	156 av:	6.86%

Table: $BPSP_2$ for test example M60-1 with $N^S = 4$.

Cycle	NS	CPU,sec.	NOC	opt NOC	Optimality gap
1	4	8.40	44	44	0.00%
2	4	6.81	68	68	0.00%
3	4	6.59	79	79	0.00%
4	4	6.64	85	85	0.00%

```
5    4    6.64    87     87      0.00%
6    4    7.75    98     91      7.69%
7    4    8.40    107    95     12.63%
8    4    6.75    112    97     15.46%
9    4    6.59    124    102    21.57%
10   4    6.64    129    111    16.22%
11   4    6.81    141    122    15.57%
12   4    6.65    147    132    11.36%
13   4    6.81    152    139     9.35%
14   4    6.92    163    149     9.40%
15   4    7.03    172    155    10.97%
------------------------------------------------------
total:    105.43   172    155   av:  8.68%
```

Table: BPSP$_2$ for test example M60-1 with $N^S = 5$.

Cycle	NS	CPU,sec.	NOC	opt NOC	Optimality gap
1	5	13.12	44	44	0.00%
2	5	8.08	68	68	0.00%
3	5	6.70	79	79	0.00%
4	5	6.54	85	85	0.00%
5	5	6.38	87	87	0.00%
6	5	7.47	94	91	3.30%
7	5	7.14	101	96	5.21%
8	5	6.70	112	98	14.29%
9	5	6.81	123	103	19.42%
10	5	7.04	129	111	16.22%
11	5	7.53	139	122	13.93%
12	5	6.86	149	132	12.88%
13	5	6.98	155	138	12.32%
14	5	9.34	164	147	11.56%
15	5	7.63	174	155	12.26%

```
------------------------------------------------------
total:    114.32   174    155   av:  8.09%
```

Finally, the results for BPSP$_3$ for test example M60-1 with $N^S = 2, ..., 5$.

Table: BPSP$_3$ for test example M60-1 with $N^S = 2$.

Cycle	NS	CPU,sec.	opt NOC
1	2	263666.65	44
2	2	3.68	68
3	2	1.32	79
4	2	1.37	86
5	2	1.26	88
6	2	1.27	92
7	2	2.42	98
8	2	2.30	103
9	2	2.75	112
10	2	1.32	120
11	2	1.43	128
12	2	1.81	134
13	2	2.52	142
14	2	1.64	152

```
15      2        2.91      161
---------------------------
total:     263694.65      161
```

Table: BPSP$_3$ for test example M60-1 with $N^S = 3$.

```
Cycle  NS  CPU,sec. opt NOC
---------------------------
1       3      9.11       44
2       3      7.03       68
3       3      7.25       79
4       3      7.08       85
5       3      6.81       87
6       3     10.00       91
7       3      8.13       95
8       3      8.07       98
9       3      8.30      105
10      3      8.18      114
11      3      9.83      125
12      3      8.18      134
13      3      7.75      140
14      3      8.79      149
15      3      8.13      156
---------------------------
total:     122.64      156
```

Table: BPSP$_3$ for test example M60-1 with $N^S = 4$.

```
Cycle  NS  CPU,sec. opt NOC
---------------------------
1       4     10.38       44
2       4      8.13       68
3       4      7.69       79
4       4      7.85       85
5       4      7.75       87
6       4      8.40       91
7       4      7.47       95
8       4      8.02       97
9       4      7.75      102
10      4     10.98      111
11      4      8.24      122
12      4      8.02      132
13      4      7.80      139
14      4     11.09      149
15      4      8.07      155
---------------------------
total:     127.64      155
```

Table: BPSP$_3$ for test example M60-1 with $N^S = 5$.

```
Cycle  NS  CPU,sec. opt NOC
---------------------------
1       5      9.06       44
2       5      8.35       68
3       5      8.40       79
4       5      8.07       85
```

```
5      5    8.08    87
6      5    7.86    91
7      5    8.74    96
8      5    8.13    98
9      5    7.96    103
10     5    8.24    111
11     5    8.41    122
12     5    8.18    132
13     5    8.30    138
14     5    8.52    147
15     5    8.18    155
---------------------------
total:     124.48   155
```

Results for BPSP$_1$ for test example M60-2 with $N^S = 2, ..., 5$

Table: BPSP$_1$ for test example M60-2 with $N^S = 2$.

Cycle	NS	CPU,sec.	NOC	opt NOC	Optimality gap
1	2	2.473	46	45	2.23%
2	2	2.574	73	64	14.06%
3	2	2.494	94	77	22.08%
4	2	2.403	110	82	34.15%
5	2	2.353	126	88	43.18%
6	2	2.493	139	90	54.44%
7	2	2.344	162	96	68.75%
8	2	2.494	179	101	77.23%
9	2	2.484	191	108	76.85%
10	2	2.293	201	118	70.34%
11	2	2.453	218	128	70.31%
12	2	2.433	236	136	73.53%
13	2	2.323	249	146	70.55%
14	2	2.273	257	152	69.08%
15	2	2.483	274	160	71.25%
total:		36.370	274	160	av: 54.53%

Table: BPSP$_1$ for test example M60-2 with $N^S = 3$.

Cycle	NS	CPU,sec.	NOC	opt NOC	Optimality gap
1	3	2.484	46	45	2.22%
2	3	2.383	75	64	17.19%
3	3	2.394	95	77	23.38%
4	3	2.293	111	82	35.37%
5	3	2.203	128	88	45.45%
6	3	2.263	145	89	62.92%
7	3	2.243	167	94	77.66%
8	3	2.233	180	97	85.57%
9	3	2.213	196	103	90.29%
10	3	2.284	206	113	82.30%
11	3	2.254	222	125	77.60%

12	3	2.243	241	132	82.58%
13	3	2.273	253	141	79.43%
14	3	2.253	260	147	76.87%
15	3	2.333	275	155	77.42%
total:		34.349	275	155	av: 61.08%

Table: BPSP$_1$ for test example M60-2 with $N^S = 4$.

Cycle	NS	CPU,sec.	NOC	opt NOC	Optimality gap
1	4	2.423	46	45	2.22%
2	4	2.223	74	64	15.63%
3	4	2.193	95	77	23.38%
4	4	2.433	112	82	36.59%
5	4	2.474	129	88	46.59%
6	4	2.353	143	89	60.67%
7	4	2.254	164	93	76.34%
8	4	2.403	179	97	84.54%
9	4	2.606	191	103	85.44%
10	4	2.366	202	112	80.36%
11	4	2.426	217	124	75.00%
12	4	2.597	231	131	76.34%
13	4	2.394	243	141	74.50%
14	4	2.284	253	146	71.61%
15	4	2.243	271	153	77.12%
total:		35.672	271	153	av: 59.06%

Table: BPSP$_1$ for test example M60-2 with $N^S = 5$.

Cycle	NS	CPU,sec.	NOC	opt NOC	Optimality gap
1	5	2.504	46	45	2.22%
2	5	2.253	75	64	17.19%
3	5	2.253	95	77	23.38%
4	5	2.263	110	82	34.15%
5	5	2.233	130	88	47.73%
6	5	2.263	143	88	62.50%
7	5	2.263	167	92	81.52%
8	5	2.214	183	96	90.63%
9	5	2.463	195	102	91.18%
10	5	2.263	206	111	85.59%
11	5	2.223	220	123	78.86%
12	5	2.463	241	132	82.58%
13	5	2.463	252	141	78.72%
14	5	2.494	264	147	79.59%
15	5	2.323	283	155	82.58%
total:		34.938	283	155	av: 62.56%

Results for BPSP$_2$ for test example M60-2 with $N^S = 2, ..., 5$

Table: BPSP$_2$ for test example M60-2 with $N^S = 2$.

Cycle	NS	CPU,sec.	NOC	opt NOC	Optimality gap
1	2	2.72	45	45	0.00%
2	2	2.63	64	64	0.00%
3	2	2.64	77	77	0.00%
4	2	2.49	82	82	0.00%
5	2	2.64	131	88	48.86%
6	2	2.46	132	90	46.67%
7	2	2.71	151	96	57.29%
8	2	2.44	162	101	60.40%
9	2	2.47	166	108	53.70%
10	2	2.65	170	118	44.07%
11	2	2.46	172	128	34.38%
12	2	2.48	217	136	59.56%
13	2	2.50	222	146	52.05%
14	2	2.47	225	152	48.03%
15	2	2.53	234	160	46.25%
total:		38.34	234	160	av: 36.75%

Table: BPSP$_2$ for test example M60-2 with $N^S = 3$.

Cycle	NS	CPU,sec.	NOC	opt NOC	Optimality gap
1	3	2.97	45	45	0.00%
2	3	3.03	64	64	0.00%
3	3	2.75	77	77	0.00%
4	3	2.81	82	82	0.00%
5	3	2.82	88	88	0.00%
6	3	2.81	126	89	41.57%
7	3	2.73	140	94	48.94%
8	3	2.75	153	97	57.73%
9	3	2.71	156	103	51.46%
10	3	2.79	167	113	47.79%
11	3	2.82	167	125	33.60%
12	3	2.74	224	132	69.70%
13	3	2.74	224	141	58.87%
14	3	2.81	227	147	54.42%
15	3	3.05	234	155	50.97%
total:		42.39	234	155	av: 34.19%

Table: BPSP$_2$ for test example M60-2 with $N^S = 4$.

Cycle	NS	CPU,sec.	NOC	opt NOC	Optimality gap
1	4	3.07	45	45	0.00%
2	4	2.94	64	64	0.00%
3	4	2.93	77	77	0.00%
4	4	2.76	82	82	0.00%
5	4	2.74	88	88	0.00%
6	4	2.76	88	89	-1.12%
7	4	2.75	133	93	43.01%
8	4	2.91	142	97	46.39%

```
9       4       2.88    149     103     44.66%
10      4       2.81    160     112     42.86%
11      4       2.79    161     124     29.84%
12      4       3.01    219     131     67.18%
13      4       2.80    224     141     58.87%
14      4       2.80    230     146     57.53%
15      4       2.79    237     153     54.90%
-----------------------------------------------------
total:          42.78   237     153  av: 29.61%
```

Table: BPSP$_2$ for test example M60-2 with $N^S = 5$.

Cycle	NS	CPU,sec.	NOC	opt NOC	Optimality gap
1	5	3.03	45	45	0.00%
2	5	2.88	64	64	0.00%
3	5	2.74	77	77	0.00%
4	5	2.69	82	82	0.00%
5	5	2.72	88	88	0.00%
6	5	2.60	88	88	3.30%
7	5	2.79	133	92	44.57%
8	5	2.73	148	96	54.17%
9	5	2.71	155	102	51.96%
10	5	2.82	163	111	46.85%
11	5	2.63	165	123	34.15%
12	5	2.77	222	132	68.18%
13	5	2.72	231	141	63.83%
14	5	2.82	231	147	57.14%
15	5	2.65	237	155	12.26%

```
-----------------------------------------------------
total:          41.36   174     155  av: 31.58%
```

Results for BPSP$_3$ for test example M60-2 with $N^S = 2, ..., 5$

Table: BPSP$_3$ for test example M60-2 with $N^S = 2$.

Cycle	NS	CPU,sec.	opt NOC
1	2	37.22	45
2	2	3.28	64
3	2	3.07	77
4	2	2.91	82
5	2	2.92	88
6	2	2.98	90
7	2	3.61	96
8	2	3.18	101
9	2	3.14	108
10	2	3.64	118
11	2	3.31	128
12	2	2.97	136
13	2	3.14	146
14	2	3.05	152
15	2	2.73	160

```
-------------------------------
total:          81.21   160
```

Table: BPSP$_3$ for test example M60-2 with $N^S = 3$.

```
Cycle  NS  CPU,sec.  opt NOC
--------------------------------
1      3    4.51       45
2      3    3.02       64
3      3    2.62       77
4      3    2.61       82
5      3    2.66       88
6      3    2.62       89
7      3    2.71       94
8      3    2.64       97
9      3    2.72      103
10     3    2.78      113
11     3    2.81      125
12     3    2.97      132
13     3    2.76      141
14     3    2.82      147
15     3    3.20      155
--------------------------------
total:      43.51     155
```

Table: BPSP$_3$ for test example M60-2 with $N^S = 4$.

```
Cycle  NS  CPU,sec.  opt NOC
--------------------------------
1      4    3.31       45
2      4    2.63       64
3      4    2.63       77
4      4    2.65       82
5      4    2.62       88
6      4    2.50       89
7      4    2.66       93
8      4    2.70       97
9      4    2.68      103
10     4    2.71      112
11     4    2.58      124
12     4    2.66      131
13     4    2.98      141
14     4    2.63      146
15     4    2.62      153
--------------------------------
total:      40.63     153
```

Table: BPSP$_3$ for test example M60-2 with $N^S = 5$.

```
Cycle  NS  CPU,sec.  opt NOC
--------------------------------
1      5    3.30       45
2      5    2.58       64
3      5    2.69       77
4      5    2.59       82
5      5    2.65       88
6      5    2.51       88
7      5    2.84       92
8      5    2.72       96
```

```
9      5    2.74    102
10     5    2.69    111
11     5    2.60    123
12     5    2.94    132
13     5    3.01    141
14     5    2.73    147
15     5    3.06    155
---------------------------
total:     41.70    155
```

Results for $BPSP_1$ for test example M60-3 with $N^S = 2, ..., 5$

Table: $BPSP_1$ for test example M60-3 with $N^S = 2$.

Cycle	NS	CPU,sec.	NOC	opt NOC	Optimality gap
1	2	2.764	46	42	9.52%
2	2	2.514	76	67	13.43%
3	2	2.583	95	75	26.67%
4	2	2.534	115	84	36.90%
5	2	2.593	135	87	55.17%
6	2	2.544	154	92	67.39%
7	2	2.563	171	97	76.29%
8	2	2.554	184	103	78.64%
9	2	2.563	194	111	74.77%
10	2	2.544	208	121	71.90%
11	2	2.554	223	130	71.54%
12	2	2.594	236	138	71.01%
13	2	2.604	246	144	70.83%
14	2	2.544	262	153	71.24%
15	2	2.614	275	159	72.96%
total:		38.666	275	159	av: 57.89%

Table: $BPSP_1$ for test example M60-3 with $N^S = 3$.

Cycle	NS	CPU,sec.	NOC	opt NOC	Optimality gap
1	3	2.384	48	42	14.29%
2	3	2.183	80	67	19.40%
3	3	2.153	97	75	29.33%
4	3	2.564	113	84	34.52%
5	3	2.784	130	86	51.16%
6	3	2.724	151	89	69.66%
7	3	2.734	168	93	80.65%
8	3	2.704	184	100	84.00%
9	3	2.824	194	106	83.02%
10	3	2.864	209	116	80.17%
11	3	2.884	227	125	81.60%
12	3	2.804	239	134	78.36%
13	3	3.025	249	137	81.75%
14	3	2.904	263	144	82.64%
15	3	2.994	276	154	79.22%
total:		40.529	276	154	av: 63.32%

Table: $BPSP_1$ for test example M60-3 with $N^S = 4$.

Cycle	NS	CPU,sec.	NOC	opt NOC	Optimality gap
1	4	2.363	45	42	7.14%
2	4	2.173	76	67	13.43%
3	4	2.123	95	75	28.00%
4	4	2.474	111	84	32.14%
5	4	2.163	129	86	50.00%
6	4	2.714	147	89	65.17%
7	4	2.724	167	93	79.57%
8	4	2.744	182	98	85.71%
9	4	2.734	190	105	80.95%
10	4	2.754	206	115	79.13%
11	4	2.964	221	123	79.67%
12	4	2.724	233	132	76.52%
13	4	2.814	240	137	75.18%
14	4	2.764	257	144	78.47%
15	4	2.885	269	153	75.82%
total:		39.117	269	153	av: 60.46%

Table: $BPSP_1$ for test example M60-3 with $N^S = 5$.

Cycle	NS	CPU,sec.	NOC	opt NOC	Optimality gap
1	5	2.423	46	42	9.52%
2	5	2.274	74	67	10.45%
3	5	2.523	90	75	20.00%
4	5	2.424	111	84	32.14%
5	5	2.744	131	86	52.33%
6	5	2.724	151	89	69.66%
7	5	2.433	164	93	76.34%
8	5	2.724	177	98	80.61%
9	5	2.965	192	104	84.62%
10	5	2.804	209	114	83.33%
11	5	2.704	226	123	83.74%
12	5	2.713	236	132	78.79%
13	5	2.753	244	136	79.41%
14	5	2.814	258	143	80.42%
15	5	2.804	271	152	78.29%
total:		39.826	271	152	av: 61.31%

Results for $BPSP_2$ for test example M60-3 with $N^S = 2, ..., 5$

Table: $BPSP_2$ for test example M60-3 with $N^S = 2$.

Cycle	NS	CPU,sec.	NOC	opt NOC	Optimality gap
1	2	2.88	42	42	0.00%
2	2	2.55	67	67	0.00%
3	2	2.55	75	75	0.00%
4	2	2.76	84	84	0.00%

Cycle	NS	CPU,sec.	NOC	opt NOC	Optimality gap
5	2	2.82	124	87	42.53%
6	2	2.42	133	92	44.57%
7	2	2.54	148	97	52.58%
8	2	2.59	155	103	50.49%
9	2	2.59	161	111	45.05%
10	2	2.56	167	121	38.02%
11	2	2.77	225	130	73.08%
12	2	2.55	226	138	63.77%
13	2	2.55	229	144	59.03%
14	2	2.47	236	153	54.25%
15	2	2.50	240	159	50.94%
total:		39.22	240	159	av: 38.29%

Table: BPSP$_2$ for test example M60-3 with $N^S = 3$.

Cycle	NS	CPU,sec.	NOC	opt NOC	Optimality gap
1	3	2.84	42	42	0.00%
2	3	2.55	67	67	0.00%
3	3	2.53	75	75	0.00%
4	3	2.53	84	84	0.00%
5	3	2.58	86	86	0.00%
6	3	2.58	132	89	48.31%
7	3	2.55	146	93	56.99%
8	3	2.57	154	100	54.00%
9	3	2.60	161	106	51.89%
10	3	2.72	166	116	43.10%
11	3	2.65	172	125	37.60%
12	3	2.57	174	134	29.85%
13	3	2.62	213	137	55.47%
14	3	2.56	222	144	54.17%
15	3	2.59	230	154	49.35%
total:		39.10	230	154	av: 32.05%

Table: BPSP$_2$ for test example M60-3 with $N^S = 4$.

Cycle	NS	CPU,sec.	NOC	opt NOC	Optimality gap
1	4	2.67	42	42	0.00%
2	4	2.73	67	67	0.00%
3	4	2.53	75	75	0.00%
4	4	2.59	84	84	0.00%
5	4	2.60	124	86	44.19%
6	4	2.54	143	89	60.67%
7	4	2.58	154	93	65.59%
8	4	2.44	159	98	62.24%
9	4	2.57	163	105	55.24%
10	4	2.57	167	115	45.22%
11	4	2.56	174	123	41.46%
12	4	2.56	174	132	31.82%
13	4	2.61	212	137	54.74%
14	4	2.57	224	144	55.56%
15	4	2.61	229	153	49.67%
total:		38.78	229	153	av: 37.76%

Table: BPSP$_2$ for test example M60-3 with $N^S = 5$.

Cycle	NS	CPU,sec.	NOC	opt NOC	Optimality gap
1	5	2.76	42	42	0.00%
2	5	2.54	67	67	0.00%
3	5	2.56	75	75	0.00%
4	5	2.57	84	84	0.00%
5	5	2.59	86	86	0.00%
6	5	2.60	132	89	48.31%
7	5	2.54	144	93	54.84%
8	5	2.56	151	98	54.08%
9	5	2.56	158	104	51.92%
10	5	2.59	163	114	42.98%
11	5	2.63	168	123	36.59%
12	5	2.57	172	132	30.30%
13	5	2.76	209	136	53.68%
14	5	2.99	223	143	55.94%
15	5	2.69	228	152	50.00%
total:		39.58	228	152	av: 31.91%

Results for BPSP$_3$ for test example M60-3 with $N^S = 2, ..., 5$

Table: BPSP$_3$ for test example M60-3 with $N^S = 2$.

Cycle	NS	CPU,sec.	opt NOC
1	2	30.44	42
2	2	6.71	67
3	2	2.62	75
4	2	2.78	84
5	2	2.68	87
6	2	3.03	92
7	2	3.01	97
8	2	3.53	103
9	2	3.15	111
10	2	3.81	121
11	2	2.88	130
12	2	2.98	138
13	2	3.29	144
14	2	3.70	153
15	2	2.73	159
total:		77.61	159

Table: BPSP$_3$ for test example M60-3 with $N^S = 3$.

Cycle	NS	CPU,sec.	opt NOC
1	3	4.05	42
2	3	2.75	67
3	3	2.75	75
4	3	2.71	84
5	3	2.71	86

6	3	2.73	89
7	3	2.74	93
8	3	2.75	100
9	3	2.55	106
10	3	3.66	116
11	3	2.90	125
12	3	2.85	134
13	3	2.99	137
14	3	2.78	144
15	3	3.00	154
total:		43.99	154

Table: BPSP$_3$ for test example M60-3 with $N^S = 4$.

Cycle	NS	CPU,sec.	opt NOC
1	4	3.38	42
2	4	2.83	67
3	4	2.84	75
4	4	2.54	84
5	4	2.88	86
6	4	2.77	89
7	4	2.79	93
8	4	2.78	98
9	4	2.61	105
10	4	2.96	115
11	4	2.80	123
12	4	2.99	132
13	4	2.70	137
14	4	2.95	144
15	4	2.91	153
total:		42.73	153

Table: BPSP$_3$ for test example M60-3 with $N^S = 5$.

Cycle	NS	CPU,sec.	opt NOC
1	5	3.44	42
2	5	2.74	67
3	5	2.54	75
4	5	2.84	84
5	5	2.75	86
6	5	2.68	89
7	5	2.77	93
8	5	2.52	98
9	5	2.88	104
10	5	2.89	114
11	5	2.80	123
12	5	2.71	132
13	5	2.85	136
14	5	2.80	143
15	5	2.91	152
total:		42.16	152

Results for $BPSP_1$ for test example M60-4 with $N^S = 2, ..., 5$

Table: $BPSP_1$ for test example M60-4 with $N^S = 2$.

Cycle	NS	CPU,sec.	NOC	opt NOC	Optimality gap
1	2	2.794	48	47	2.13%
2	2	2.754	74	67	10.45%
3	2	2.904	98	83	18.07%
4	2	2.814	115	85	35.29%
5	2	2.925	131	91	43.96%
6	2	2.754	146	94	55.32%
7	2	2.854	161	98	64.29%
8	2	3.014	174	104	67.31%
9	2	3.085	191	109	75.23%
10	2	2.934	205	119	72.27%
11	2	2.974	219	125	75.20%
12	2	2.894	229	135	69.63%
13	2	3.034	245	139	76.26%
14	2	2.904	259	151	71.52%
15	2	2.944	271	160	69.38%
total:		43.582	271	160	av: 53.75%

Table: $BPSP_1$ for test example M60-4 with $N^S = 3$.

Cycle	NS	CPU,sec.	NOC	opt NOC	Optimality gap
1	3	2.744	48	47	2.13%
2	3	2.724	75	67	11.94%
3	3	2.324	97	83	16.87%
4	3	2.533	115	85	35.29%
5	3	2.294	132	89	48.31%
6	3	2.383	147	92	59.78%
7	3	2.223	165	95	73.68%
8	3	2.233	176	99	77.78%
9	3	2.213	194	107	81.31%
10	3	2.243	208	116	79.31%
11	3	2.184	223	123	81.30%
12	3	2.293	233	129	80.62%
13	3	2.213	249	137	81.75%
14	3	2.253	262	148	77.03%
15	3	2.394	274	154	77.92%
total:		35.251	274	154	av: 59.00%

Table: $BPSP_1$ for test example M60-4 with $N^S = 4$.

Cycle	NS	CPU,sec.	NOC	opt NOC	Optimality gap
1	4	2.794	48	47	2.13%
2	4	2.754	75	67	11.94%
3	4	2.784	100	83	20.48%

```
4     4    2.895    116      85      36.47%
5     4    2.864    135      89      51.69%
6     4    2.884    151      92      64.13%
7     4    2.874    170      94      80.85%
8     4    2.624    179      99      80.81%
9     4    2.794    196     105      86.67%
10    4    2.834    210     116      81.03%
11    4    2.744    223     123      81.30%
12    4    2.704    233     129      80.62%
13    4    2.604    251     134      87.31%
14    4    2.724    265     146      81.51%
15    4    2.804    278     153      81.70%
-------------------------------------------------
total:    41.681    278     153  av: 61.91%
```

Table: BPSP$_1$ for test example M60-4 with $N^S = 5$.

Cycle	NS	CPU,sec.	NOC	opt NOC	Optimality gap
1	5	2.464	48	47	2.13%
2	5	2.693	75	67	11.94%
3	5	2.223	100	83	20.48%
4	5	2.283	116	85	36.47%
5	5	2.273	136	89	52.81%
6	5	2.164	152	92	65.22%
7	5	2.153	167	94	77.66%
8	5	2.163	174	99	75.76%
9	5	2.303	188	105	79.05%
10	5	2.183	203	115	76.52%
11	5	2.194	212	122	73.77%
12	5	2.213	222	128	73.44%
13	5	2.113	242	133	81.95%
14	5	2.343	256	146	75.34%
15	5	2.294	269	152	76.97%
total:		34.059	269	152	av: 58.63%

Results for BPSP$_2$ for test example M60-4 with $N^S = 2, ..., 5$

Table: BPSP$_2$ for test example M60-4 with $N^S = 2$.

Cycle	NS	CPU,sec.	NOC	opt NOC	Optimality gap
1	2	2.76	47	47	0.00%
2	2	2.63	67	67	0.00%
3	2	2.61	83	83	0.00%
4	2	2.58	122	85	43.53%
5	2	2.75	135	91	48.35%
6	2	2.89	146	94	55.32%
7	2	3.03	155	98	58.16%
8	2	3.01	161	104	54.81%
9	2	3.08	166	109	52.29%
10	2	3.06	171	119	43.70%
11	2	3.22	173	125	38.40%
12	2	2.94	173	135	28.15%

```
13      2      2.85     178      139        28.06%
14      2      3.01     217      151        43.71%
15      2      3.22     229      160        43.13%
------------------------------------------------------
total:         43.68    229      160  av: 35.84%
```

Table: BPSP$_2$ for test example M60-4 with $N^S = 3$.

Cycle	NS	CPU,sec.	NOC	opt NOC	Optimality gap
1	3	3.07	47	47	0.00%
2	3	3.00	67	67	0.00%
3	3	3.16	83	83	0.00%
4	3	2.99	85	85	0.00%
5	3	2.98	132	89	48.31%
6	3	3.10	136	92	47.83%
7	3	2.97	148	95	55.79%
8	3	2.93	159	99	60.61%
9	3	2.96	164	107	53.27%
10	3	2.82	169	116	45.69%
11	3	3.04	171	123	39.02%
12	3	2.76	171	129	32.56%
13	3	3.02	175	137	27.74%
14	3	2.98	218	148	47.30%
15	3	3.10	224	154	45.45%

```
------------------------------------------------------
total:         44.93    224      154  av: 33.57%
```

Table: BPSP$_2$ for test example M60-4 with $N^S = 4$.

Cycle	NS	CPU,sec.	NOC	opt NOC	Optimality gap
1	4	2.84	47	47	0.00%
2	4	2.55	67	67	0.00%
3	4	2.90	83	83	0.00%
4	4	3.09	85	85	0.00%
5	4	3.12	132	89	48.31%
6	4	3.11	136	92	47.83%
7	4	3.04	146	94	55.32%
8	4	3.06	159	99	60.61%
9	4	3.08	159	105	51.43%
10	4	3.12	167	116	43.97%
11	4	3.07	167	123	35.77%
12	4	3.03	167	129	29.46%
13	4	3.04	173	134	29.10%
14	4	3.02	215	146	47.26%
15	4	3.05	219	153	43.14%

```
------------------------------------------------------
total:         45.16    219      153  av: 32.69%
```

Table: BPSP$_2$ for test example M60-4 with $N^S = 5$.

Cycle	NS	CPU,sec.	NOC	opt NOC	Optimality gap
1	5	2.83	47	47	0.00%
2	5	2.99	67	67	0.00%

```
3     5    3.07   83     83    0.00%
4     5    2.80   85     85    0.00%
5     5    3.00  132     89   48.31%
6     5    2.84  145     92   57.61%
7     5    2.70  152     94   61.70%
8     5    3.15  162     99   63.64%
9     5    3.20  164    105   56.19%
10    5    3.08  172    115   49.57%
11    5    3.16  173    122   41.80%
12    5    3.19  171    128   33.59%
13    5    3.20  176    133   32.33%
14    5    3.01  216    146   47.95%
15    5    3.18  225    152   48.03%
------------------------------------------------------
total:    45.45  225    152  av: 36.05%
```

Results for BPSP$_3$ for test example M60-4 with $N^S = 2, ..., 5$

Table: BPSP$_3$ for test example M60-4 with $N^S = 2$.

```
Cycle  NS  CPU,sec.  opt NOC
-----------------------------
1      2  218974.44      47
2      2       3.84      67
3      2       3.07      83
4      2       3.04      85
5      2       3.26      91
6      2       3.13      94
7      2       3.46      98
8      2       3.22     104
9      2       3.37     109
10     2       3.74     119
11     2       3.26     125
12     2       3.88     135
13     2       3.24     139
14     2       4.58     151
15     2       3.86     160
-----------------------------
total:    219023.43     160
```

Table: BPSP$_3$ for test example M60-4 with $N^S = 3$.

```
Cycle  NS  CPU,sec.  opt NOC
-----------------------------
1      3     4.31        47
2      3     3.13        67
3      3     3.09        83
4      3     3.12        85
5      3     3.17        89
6      3     3.04        92
7      3     3.08        95
8      3     3.61        99
9      3     3.79       107
10     3     3.56       116
11     3     3.19       123
```

```
12      3    4.18      129
13      3    3.71      137
14      3    3.74      148
15      3    3.23      154
---------------------------
total:       51.98     154
```

Table: BPSP$_3$ for test example M60-4 with $N^S = 4$.

```
Cycle  NS   CPU,sec. opt NOC
---------------------------
1       4    3.58       47
2       4    3.37       67
3       4    3.24       83
4       4    3.19       85
5       4    3.49       89
6       4    3.04       92
7       4    3.31       94
8       4    3.67       99
9       4    3.60      105
10      4    3.33      116
11      4    3.17      123
12      4    3.42      129
13      4    3.41      134
14      4    3.21      146
15      4    3.32      153
---------------------------
total:       50.36     153
```

Table: BPSP$_3$ for test example M60-4 with $N^S = 5$.

```
Cycle  NS   CPU,sec. opt NOC
---------------------------
1       5    3.60       47
2       5    3.33       67
3       5    3.32       83
4       5    3.43       85
5       5    3.15       89
6       5    3.05       92
7       5    3.14       94
8       5    3.11       99
9       5    3.19      105
10      5    3.16      115
11      5    3.14      122
12      5    3.18      128
13      5    3.17      133
14      5    3.35      146
15      5    3.19      152
---------------------------
total:       48.55     152
```

Results for BPSP$_1$ for test example M60-5 with $N^S = 2, ..., 5$

Table: BPSP$_1$ for test example M60-5 with $N^S = 2$.

```
Cycle  NS   CPU,sec.    NOC    opt NOC   Optimality gap
```

```
1       2    2.554    46     41     12.20%
2       2    2.694    78     67     16.42%
3       2    2.764   100     77     29.87%
4       2    2.874   119     86     38.37%
5       2    2.874   136     90     51.11%
6       2    2.864   152     93     63.44%
7       2    2.914   164     97     69.07%
8       2    2.774   183    104     75.96%
9       2    2.413   197    111     77.48%
10      2    2.253   212    119     78.15%
11      2    2.834   223    129     72.87%
12      2    2.825   241    137     75.91%
13      2    2.794   254    145     75.17%
14      2    2.904   263    153     71.90%
15      2    2.824   279    160     74.38%
```
```
total:      41.159   279    160  av: 58.82%
```

Table: BPSP$_1$ for test example M60-5 with $N^S = 3$.

Cycle	NS	CPU,sec.	NOC	opt NOC	Optimality gap
1	3	2.674	46	41	12.20%
2	3	2.684	76	67	13.43%
3	3	2.363	95	77	23.38%
4	3	2.724	116	85	36.47%
5	3	2.864	133	89	49.44%
6	3	2.244	151	91	65.93%
7	3	3.074	163	95	71.58%
8	3	2.674	180	98	83.67%
9	3	2.594	196	105	86.67%
10	3	2.253	210	114	84.21%
11	3	2.524	219	122	79.51%
12	3	2.413	235	129	82.17%
13	3	2.824	249	141	76.60%
14	3	2.955	260	148	75.68%
15	3	3.124	275	157	75.16%

```
total:      39.988   275    157  av: 61.07%
```

Table: BPSP$_1$ for test example M60-5 with $N^S = 4$.

Cycle	NS	CPU,sec.	NOC	opt NOC	Optimality gap
1	4	2.583	45	41	9.76%
2	4	2.264	76	67	13.43%
3	4	2.804	93	77	20.78%
4	4	2.714	114	85	34.12%
5	4	2.824	133	89	49.44%
6	4	2.834	152	91	67.03%
7	4	2.834	164	94	74.47%
8	4	2.513	183	97	88.66%
9	4	2.804	198	103	92.23%
10	4	2.975	212	113	87.61%
11	4	2.924	223	121	84.30%

12	4	2.283	241	126	91.27%
13	4	2.804	254	138	84.06%
14	4	3.125	266	146	82.19%
15	4	2.934	278	156	78.21%
total:		41.629	278	156	av: 63.84%

Table: BPSP$_1$ for test example M60-5 with $N^S = 5$.

Cycle	NS	CPU,sec.	NOC	opt NOC	Optimality gap
1	5	2.543	44	41	7.32%
2	5	2.614	74	67	10.45%
3	5	2.774	97	77	25.97%
4	5	2.714	118	85	38.82%
5	5	2.424	136	89	52.81%
6	5	2.674	151	91	65.93%
7	5	2.393	161	94	71.28%
8	5	2.123	180	98	83.67%
9	5	2.864	191	103	85.44%
10	5	2.965	206	113	82.30%
11	5	2.924	218	121	80.17%
12	5	2.804	239	126	89.68%
13	5	3.004	250	138	81.16%
14	5	2.934	260	146	78.08%
15	5	2.874	270	156	73.08%
total:		40.628	270	156	av: 61.74%

Results for BPSP$_2$ with $N^S = 2, ..., 5$ for test example M60-5

Table: BPSP$_2$ for test example M60-5 with $N^S = 2$.

Cycle	NS	CPU,sec.	NOC	opt NOC	Optimality gap
1	2	2.67	41	41	0.00%
2	2	2.85	67	67	0.00%
3	2	2.55	77	77	0.00%
4	2	3.12	85	86	-1.16%
5	2	3.14	130	90	44.44%
6	2	2.96	132	93	41.94%
7	2	3.12	142	97	46.39%
8	2	3.07	150	104	44.23%
9	2	3.09	162	111	45.95%
10	2	3.01	167	119	40.34%
11	2	3.02	169	129	31.01%
12	2	3.08	219	137	59.85%
13	2	2.99	223	145	53.79%
14	2	3.03	229	153	49.67%
15	2	3.17	234	160	46.25%
total:		45.33	234	160	av: 33.51%

Table: BPSP$_2$ for test example M60-5 with $N^S = 3$.

Cycle	NS	CPU,sec.	NOC	opt NOC	Optimality gap
1	3	2.81	41	41	0.00%
2	3	2.55	67	67	0.00%
3	3	2.53	77	77	0.00%
4	3	2.53	85	85	0.00%
5	3	3.13	131	89	47.19%
6	3	3.13	132	91	45.05%
7	3	3.07	144	95	51.58%
8	3	3.34	150	98	53.06%
9	3	3.26	162	105	54.29%
10	3	3.18	168	114	47.37%
11	3	3.22	172	122	40.98%
12	3	3.04	175	129	35.66%
13	3	3.11	217	141	53.90%
14	3	3.20	220	148	48.65%
15	3	3.46	229	157	45.86%
total:		46.29	229	157	av: 34.63%

Table: BPSP$_2$ for test example M60-5 with $N^S = 4$.

Cycle	NS	CPU,sec.	NOC	opt NOC	Optimality gap
1	4	2.63	41	41	0.00%
2	4	2.58	67	67	0.00%
3	4	2.63	77	77	0.00%
4	4	2.45	85	85	0.00%
5	4	2.49	131	89	47.19%
6	4	2.43	136	91	49.45%
7	4	2.52	147	94	56.38%
8	4	2.44	150	97	54.64%
9	4	2.61	164	103	59.22%
10	4	2.39	165	113	46.02%
11	4	2.36	167	121	38.02%
12	4	2.45	175	126	38.89%
13	4	2.68	216	138	56.52%
14	4	2.46	223	146	52.74%
15	4	2.43	231	156	48.08%
total:		37.57	231	156	av: 36.48%

Table: BPSP$_2$ for test example M60-5 with $N^S = 5$.

Cycle	NS	CPU,sec.	NOC	opt NOC	Optimality gap
1	5	2.81	41	41	0.00%
2	5	2.58	67	67	0.00%
3	5	2.61	77	77	0.00%
4	5	2.76	85	85	0.00%
5	5	2.88	131	89	47.19%
6	5	3.16	133	91	46.15%
7	5	3.00	143	94	52.13%
8	5	2.90	152	98	55.10%
9	5	3.05	162	103	57.28%
10	5	3.02	165	113	46.02%

```
11    5    3.02    169    121    39.67%
12    5    3.06    173    126    37.30%
13    5    3.05    216    138    56.52%
14    5    3.04    217    146    48.63%
15    5    3.14    225    156    44.23%
------------------------------------------------
total:      44.12   225    156  av: 35.35%
```

Results for BPSP$_3$ for test example M60-5 with $N^S = 2, ..., 5$

Table: BPSP$_3$ for test example M60-5 with $N^S = 2$.

```
Cycle  NS  CPU,sec.  opt NOC
---------------------------------
1      2   26721.14      41
2      2       4.24      67
3      2       3.07      77
4      2       3.36      86
5      2       3.16      90
6      2       3.18      93
7      2       3.08      97
8      2       3.30     104
9      2       3.40     111
10     2       3.41     119
11     2       3.46     129
12     2       3.48     137
13     2       3.56     145
14     2       3.89     153
15     2       3.22     160
---------------------------------
total:     26768.97     160
```

Table: BPSP$_3$ for test example M60-5 with $N^S = 3$.

```
Cycle  NS  CPU,sec.  opt NOC
---------------------------------
1      3    4.29      41
2      3    3.10      67
3      3    3.22      77
4      3    3.18      85
5      3    3.18      89
6      3    3.24      91
7      3    3.38      95
8      3    3.27      98
9      3    3.78     105
10     3    3.41     114
11     3    3.30     122
12     3    3.70     129
13     3    3.59     141
14     3    3.47     148
15     3    3.43     157
---------------------------------
```

```
total:      51.56     157
```

Table: BPSP$_3$ for test example M60-5 with $N^S = 4$.

```
Cycle  NS  CPU,sec. opt NOC
----------------------------
1      4    3.73        41
2      4    3.13        67
3      4    3.24        77
4      4    3.19        85
5      4    3.36        89
6      4    3.12        91
7      4    3.20        94
8      4    3.38        97
9      4    3.36       103
10     4    3.40       113
11     4    3.43       121
12     4    3.17       126
13     4    3.67       138
14     4    3.26       146
15     4    3.15       156
----------------------------
total:     49.83       156
```

Table: BPSP$_3$ for test example M60-5 with $N^S = 5$.

```
Cycle  NS  CPU,sec. opt NOC
----------------------------
1      5    3.56        41
2      5    3.04        67
3      5    3.26        77
4      5    3.16        85
5      5    3.24        89
6      5    3.09        91
7      5    3.21        94
8      5    3.16        98
9      5    3.20       103
10     5    3.36       113
11     5    3.26       121
12     5    3.42       126
13     5    3.20       138
14     5    3.80       146
15     5    3.47       156
----------------------------
total:     49.45       156
```

Results for combination of BPSP$_2$ and BPSP$_3$ for test example M60-4 with $N^S = 2$
(cycle 1: BPSP$_2$; cycle 2-14: BPSP$_3$)

```
Cycle  NS   CPU,sec.   NOC
----------------------------
1      2     2.76       47
2      2     3.05       67
3      2     2.92       83
4      2     2.52       85
```

```
5      2      3.27      90
6      2      3.80      95
7      2      2.98      97
8      2      3.04     104
9      2      3.49     113
10     2      3.47     122
11     2      3.21     128
12     2      3.39     133
13     2      4.07     143
14     2      4.21     153
15     2      4.18     161
-----------------------------
total:        50.38    161
```

Results for combination of BPSP$_2$ and BPSP$_3$ for test example M60-5 with $N^S = 2$ (cycle 1: BPSP$_2$; cycle 2-14: BPSP$_3$)

```
Cycle  NS   CPU,sec.   NOC
-----------------------------
1      2      3.07      41
2      2      3.71      67
3      2      3.09      77
4      2      3.39      85
5      2      3.02      89
6      2      3.05      91
7      2      3.26      97
8      2      3.18     103
9      2      3.30     111
10     2      3.96     120
11     2      3.46     132
12     2      2.70     136
13     2      3.07     145
14     2      3.52     156
15     2      3.76     163
-----------------------------
total:        49.57    163
```

As one can see applying combinations of the models produce near optimal ($\Delta \approx$ 0.1% − 0.2%) solutions ($\Delta \approx 0.1\% - 0.2\%$) much faster than using BPSP$_3$ only.

A.2.   Tabular Results for Different Algorithms

In the following tables and figures the numerical results for the algorithms D_2, WL_1, WL_n^1, WL_n^2, SL_n are presented. Note that the CPU time is not reported because it is always was less than 6 seconds. The figures do not show the results obtained using the algorithm SL_n as those values cannot easily displayed on the same scale.

n	D_2	WL_1	WL_n^1	WL_n^2	SL_n
1	484	480	484	484	549
2	484	480	489	489	548
3	484	480	491	490	547
4	484	480	486	485	544
5	484	480	484	483	550

Table 1: Results with lower bound 472

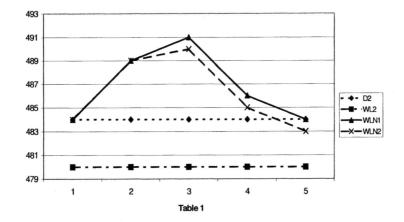

Table 1

n	D_2	WL_1	WL_n^1	WL_n^2	SL_n
1	484	485	483	483	562
2	484	485	484	484	559
3	484	485	485	485	557
4	484	485	482	482	551
5	484	485	482	482	553

Table 2: Results with lower bound 475

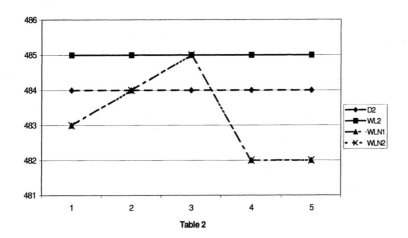

Table 2

n	D_2	WL_1	WL_n^1	WL_n^2	SL_n
1	490	489	487	484	562
2	490	489	487	487	564
3	490	489	485	484	561
4	490	489	488	487	561
5	490	489	485	484	556

Table 3: Results with lower bound 478

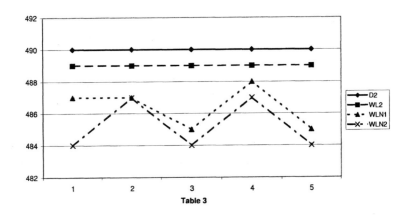

n	D_2	WL_1	WL_n^1	WL_n^2	SL_n
1	483	479	478	477	581
2	483	479	479	476	579
3	483	479	481	479	579
4	483	479	483	481	572
5	483	479	481	479	565

Table 4: Results with lower bound 470

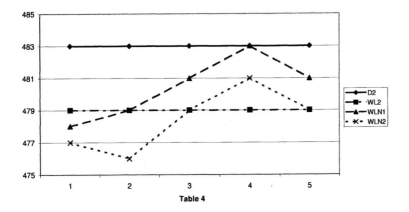

Table 4

n	D_2	WL_1	WL_n^1	WL_n^2	SL_n
1	475	477	482	481	552
2	475	477	483	482	556
3	475	477	481	482	560
4	475	477	483	480	558
5	475	477	484	482	559

Table 5: Results with lower bound 472

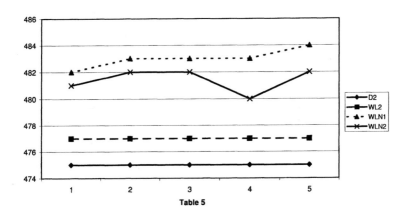

Table 5

n	D_2	WL_1	WL_n^1	WL_n^2	SL_n
1	482	482	485	485	562
2	482	482	488	487	562
3	482	482	487	486	562
4	482	482	486	486	559
5	482	482	485	485	557

Table 6: Results with lower bound 475

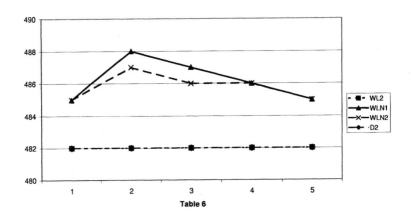

n	D_2	WL_1	WL_n^1	WL_n^2	SL_n
1	479	476	477	477	564
2	479	476	481	480	563
3	479	476	482	481	556
4	479	476	479	478	559
5	479	476	476	475	561

Table 7: Results with lower bound 467

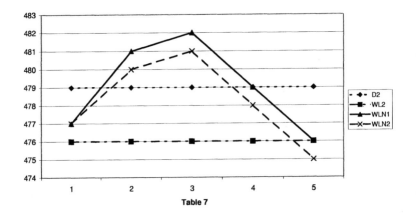

n	D_2	WL_1	WL_n^1	WL_n^2	SL_n
1	482	483	486	486	551
2	482	483	485	485	550
3	482	483	488	488	544
4	482	483	488	488	547
5	482	483	487	487	544

Table 8: Results with lower bound 473

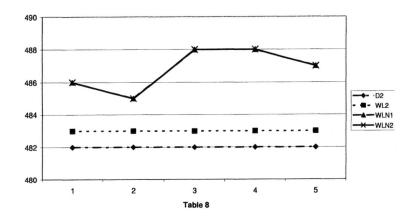

n	D_2	WL_1	WL_n^1	WL_n^2	SL_n
1	485	484	491	489	565
2	485	484	490	488	565
3	485	484	492	488	560
4	485	484	491	487	559
5	485	484	494	489	557

Table 9: Results with lower bound 478

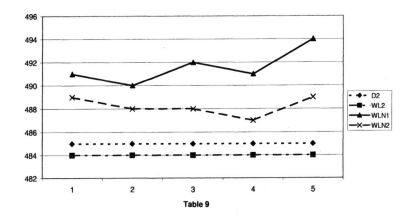

n	D_2	WL_1	WL_n^1	WL_n^2	SL_n
1	479	478	480	478	548
2	479	478	484	482	547
3	479	478	482	479	543
4	479	478	485	482	551
5	479	478	485	482	550

Table 10: Results with lower bound 472

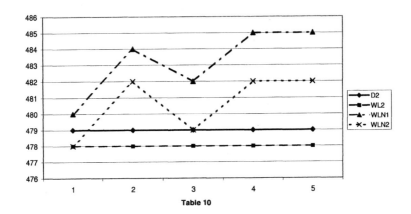

Appendix B
OptiTrans

B.1. Input Data

B.1.1 Input Data Common to all Solution Approaches

Our implementation requires the following order-specific information displayed in Fig. B.1.1:

1 The name of a pick-up order;

2 The name of a delivery order;

3 The name of the pick-up building;

4 The physical node of the pick-up building;

5 The name of the delivery building;

6 The physical node of the delivery building;

7 Desired pick-up time (if no time specified, then we set this time to zero and do not consider any pick-up time);

8 Desired delivery time (if no time specified, then we set this time to zero and do not consider any pick-up time);

9 The type of an order (*i.e.*, whether a vehicle transport sitting, lying patient or some materials and so on...);

10 A flag, indicating whether a patient is infectious;

11 The number of accompanied persons;

12 The service time required for picking-up a patient;

13 The service time required for delivering a patient;

14 The number of the vehicle, if the patient is already in that vehicle (if an order still need to be served, then zero);

15 In case the previous parameter is not zero, then this is the amount of minutes a patient is in a car already.

14

A25524	25524	RA-A	87	M5-01	60	13.5	0	2	0	0	0	6	4	0
A25501	25501	A-04	8	M2-A	86	13.75	0	2	0	0	9	3	5	0
A25504	25504	A-02	8	M2-A	86	13.75	0	2	0	0	9	3	4	0
A25528	25528	M3-KARD	71	M3-02	80	13.766	0	1	0	0	6	9	0	0
A25525	25525	M3-OP	80	M5-03	60	14	0	1	0	0	3	6	2	0
A25531	25531	M3-08	71	M3-OP	80	13.85	0	2	0	0	6	3	5	0
A25532	25532	NN-A	55	CM-A	69	13.888	0	1	0	0	3	3	3	0
A25527	25527	M5-02	60	RD-CT57	69	14	0	1	0	0	9	3	0	0
A25508	25508	HO-02	5	GK-01	89	14	0	2	0	0	22	5	6	0
A25441	25441	HO-01	5	M2-A	86	14	0	2	0	0	21	3	7	0
A25364	25364	HO-03	5	RA-A	87	14.45	14.5	1	0	0	30	9	0	0
A25530	25530	D-03	21	RD-R0E49	86	14	0	2	0	0	3	3	0	0
A25533	25533	M3-OP	80	M5-03	60	14.5	0	1	0	0	3	6	0	0
A25534	25534	M5-03	60	M3-OP	80	14.25	0	1	0	0	6	3	0	0

Figure B.1.1.   Input file: order-dependent information

Column nine might indicate :

- 1 - the patient is in bed;

- 2 - regular patient;

- 3 - the patient is in a wheel-chair;

- 4 - materials should be transported;

- 10 - break for the car, it means the car has to drive to the depo and to stay there for the time specified in columns 12 and 13.

If the value in column nine is negative it means that corresponding order is urgent. For offline cases the columns 14 and 15 are neglected, because all values in those are zero.

Fig. B.1.2 contains the vehicle-dependent information:

- The total number of vehicles and the current time in seconds;

    Starting from the second line, the content of the column:

- The number of the vehicle;

- Current location of the vehicle (number of the building);

- Earliest time when the vehicle is available in seconds;

- Latest time when the vehicle is available in seconds;

- The number of patients in bed the vehicle can transport;

- The number of sitting patients the vehicle can transport;

- The wheel-chair capacity of the vehicle.

In addition, all control parameters specified in Section B.B.1.3.2 are required.

B.1.2   Specific Input Data for the MILP Model and the Column Enumeration Approach

The MILP models involves several parameters used in the fine tuning process generating cuts:

12	50861					
1	80	50862	86399	2	5	1
2	80	50861	51120	2	5	1
3	54	50861	51120	2	5	1
4	87	50861	54720	2	5	1
5	34	50861	56520	2	5	1
6	5	50861	57600	2	5	1
7	5	50861	57600	2	5	1
8	69	50861	58320	2	5	1
9	55	50861	58320	2	5	1
10	11	50861	58320	2	5	1
11	69	50861	60120	2	5	1
12	73	50861	60120	2	5	1

Figure B.1.2. Input file: vehicle-dependent information

- Δ^B 2.0 - used to reduce the arc system (backward) seen by the MILP model,

- Δ^F 2.0 - used to reduce the arc system (forward) seen by the MILP model,

- R^{A_1} 2.0 - the maximum allowed deviations (earliness) from target time for pickup nodes,

- R^{A_2} 2.0 - the maximum allowed deviations (lateness) from target time for pickup nodes,

- R^B 2.0 - the maximum allowed deviations (lateness) from target time for delivery nodes,

- S_L^+ 2.0 - the maximum allowed sum of lateness, *e.g.*, derived from the heuristic.

B.1.3 Specific Input Data for the Heuristic Methods

B.1.3.1 Penalty Criteria

We use several parameters to penalize violations of some constraints:

- **PENINF1** 5555.0 - used to penalize capacity infeasibility (*i.e.*, an order is assigned to a vehicle v for which either $c_j^L = C_v^L$, $c_j^W = C_v^W$ or $c_j^S = C_v^S$).

- **PENPAR1** 3001.0 - used when an order is not in vehicle's time windows (*i.e.*, an order is assigned to a vehicle which is "not yet" or "already not" available, see Fig. B.1.2).

- **PENPAR2** 1000.0 - used when a pre-assigned order is arrived too late.

- **PENPAR3** 200.0 - used to avoid that a patient is in a vehicle for more than $\Delta T = 40$ minutes. All what is more than 40 minutes is penalized.

- **PENPAR4** 444.0 - used to enforce the desired time when a vehicle should take its lunch break.

B.1.3.2 Control Parameters of the OptiTrans Software

In this section we provide a detailed list of the control parameters used in our construction and improvement heuristics.

- IFDRO is used to decide how remaining orders should be distributed.

  1 IFDRO 0 implies that remaining orders are neglected;

  2 IFDRO 1 implies that remaining orders are distributed using DU heuristic;

  3 IFDRO 2 implies that remaining orders are distributed using DB heuristic.

- IFRDSTR is used to choose the imprrovement heuristic.

  1 IFRDSTR 0 implies that IT1 heuristic is used;

  2 IFRDSTR 1 implies that the combination of IT1 and IT2 is used with first acceptance criteria (see p.140);

  3 IFRDSTR 2 implies that the combination of IT1 and IT2 is used with second acceptance criteria (see p.140);

  4 IFRDSTR 3 implies that IT2 heuristic is used.

- INCOH is used to choose the construction heuristic

  1 INCOH 1 implies that remaining orders are distributed using CONH1 heuristic;

  2 INCOH 2 implies that remaining orders are distributed using CONH2 heuristic.

- IOPTMD is used to select either offline or online optimization.

  1 IOPTMD 1 selects offline optimization;

  2 IOPTMD 2 selects online optimization.

- PXINC 0.05 is the permitted increase of the maximal lateness in accepting a tour to take an order in the cross-over scheme. This tuning parameter is one of the most sensitive parameters in the heuristic. Its value is recommend to be chosen as $0.15D_v$ to $0.5D_v$.

- TIMECT 125.0 defines the cleaning time after transporting an infectious person.

- TIMECC1 5.0 defines the time in minutes for changing a car at the depot.

- TIMECC2 5.0 defines the time in minutes for changing a car at the depot.

- TOLO 0.058 sets the tolerance on accepting two orders having the same pick-up time for the same vehicle.

- TOL1 0.25 sets the tolerance on accepting the lower limit of the vehicle's availability.

- TOL2 0.25 sets the tolerance on accepting the upper limit of the vehicle's availability.

B.2. Tabular Results
B.2.1 Tabular Results for the MILP Model

Case M01: optimal tour for the set of 4 orders

v	node	TE	tA	tL	rA2	rB	S	L
2	A150826	7.500	7.500		–	–	0	1
2	A150752	7.500	7.655		0.154500	–	1	1
2	150826		7.781	8.000	–	–	1	0
2	150752		7.860	8.000	–	–	0	0
2	A150869	8.000	8.000		–	–	1	0
2	A150890	8.000	8.025		0.025133	–	2	0
2	150869		8.334	8.500	–	–	1	0
2	150890		8.392	8.500	–	–	0	0

Case M02: optimal tour for the set of 5 orders

v	node	TE	tA	tL	rA2	rB	S	L
2	A150826	7.500	7.500		–	–	0	1
2	A150752	7.500	7.655		0.154500	–	1	1
2	150826		7.781	8.000	–	–	1	0
2	150752		7.860	8.000	–	–	0	0
2	A150869	8.000	8.000		–	–	1	0
2	A150890	8.000	8.025		0.025133	–	2	0
2	150890		8.322	8.500	–	–	1	0
2	150869		8.418	8.500	–	–	0	0
2	A150922	8.583	8.583		–	–	0	1
2	150922		8.828	9.083	–	–	0	0

Case M03: optimal tour for the set of 6 orders

v	node	TE	tA	tL	rA2	rB	S	L
2	A150826	7.500	7.500		–	–	0	1
2	A150752	7.500	7.655		0.154500	–	1	1
2	150826		7.781	8.000	–	–	1	0

v	node	TE	tA	tL	rA2	rB	S	L
2	150752		7.860	8.000	–	–	0	0
2	A150869	8.000	8.000		–	–	1	0
2	A150890	8.000	8.025		0.025133	–	2	0
2	150890		8.322	8.500	–	–	1	0
2	150869		8.418	8.500	–	–	0	0
2	A150922	8.583	8.583		–	–	0	1
2	150922		8.828	9.083	–	–	0	0
2	A150920	9.000	9.015		0.014763	–	0	1
2	150920		9.322	9.500	–	–	0	0

Case M04: optimal tour for the set of 7 orders

v	node	TE	tA	tL	rA2	rB	S	L
2	A150826	7.500	7.500		–	–	0	1
2	A150752	7.500	7.655		0.154500	–	1	1
2	150826		7.781	8.000	–	–	1	0
2	150752		7.860	8.000	–	–	0	0
2	A150869	8.000	8.000		–	–	1	0
2	A150890	8.000	8.025		0.025133	–	2	0
2	150890		8.322	8.500	–	–	1	0
2	150869		8.418	8.500	–	–	0	0
2	A150922	8.583	8.583		–	–	0	1
2	150922		8.828	9.083	–	–	0	0
2	A150920	9.000	9.015		0.014763	–	0	1
2	150920		9.322	9.500	–	–	0	0
2	A150939	9.500	9.500		–	–	1	0
2	150939		9.548	10.000	–	–	0	0

Case M05: optimal tour for the set of 13 orders

v	node	TE	tA	tL	rA2	rB	S	L
1	A150826	7.500	7.500		–	–	0	1
1	A150752	7.500	7.655		0.154500	–	1	1
1	150826		7.781	8.000	–	–	1	0
1	150752		7.860	8.000	–	–	0	0
1	A150869	8.000	8.000		–	–	1	0
1	A150890	8.000	8.025		0.025133	–	2	0
1	150869		8.334	8.500	–	–	1	0
1	150890		8.500	8.500	–	–	0	0

v	node	TE	tA	tL	rA2	rB	S L
1	A150922	8.583	8.583		–	–	0 1
1	150922		8.828	9.083	–	–	0 0
1	A150920	9.000	9.015		0.014763	–	0 1
1	150920		9.500	9.500	–	–	0 0
1	A150977	10.250	10.250		–	–	0 1
1	150977		10.525	10.750	–	–	0 0
1	A150749	11.000	11.000		–	–	1 0
1	150749		11.248	11.500	–	–	0 0
1	A151000	11.250	11.250		–	–	0 1
1	A150933	11.500	11.500		–	–	1 1
1	151000		11.804	11.750	–	0.053600	1 0
1	150933		12.000	12.000	–	–	0 0
1	A151038	12.517	12.517		–	–	0 1
1	151038		12.900	13.017	–	–	0 0
1	A151022	13.250	13.250		–	–	0 1
1	151022		13.442	13.750	–	–	0 0

Case M06: optimal tour for the whole set of 16 orders

v	node	TE	tA	tL	rA2	rB	S L
2	A150826	7.500	7.500		–	–	0 1
2	A150752	7.500	7.655		0.154500	–	1 1
2	150826		7.781	8.000	–	–	1 0
2	150752		7.860	8.000	–	–	0 0
2	A150869	8.000	8.000		–	–	1 0
2	A150890	8.000	8.025		0.025133	–	2 0
2	150890		8.322	8.500	–	–	1 0
2	150869		8.418	8.500	–	–	0 0
2	A150922	8.583	8.583		–	–	0 1
2	150922		8.828	9.083	–	–	0 0
2	A150920	9.000	9.015		0.014763	–	0 1
2	150920		9.322	9.500	–	–	0 0
2	A150939	9.500	9.500		–	–	1 0
2	150939		9.548	10.000	–	–	0 0
2	A150946	9.600	9.600		–	–	0 1
2	150946		9.807	10.100	–	–	0 0
2	A150965	9.983	9.983		–	–	0 1
2	150965		10.082	10.483	–	–	0 0
2	A150970	10.250	10.250		–	–	1 0
2	A150977	10.250	10.284		0.034267	–	1 1
2	150970		10.611	10.750	–	–	0 1
2	150977		10.635	10.750	–	–	0 0
2	A150749	11.000	11.000		–	–	1 0
2	A151000	11.250	11.250		–	–	1 1
2	150749		11.402	11.500	–	–	0 1

2	A150933	11.500	11.500		–	–	1	1
2	151000		11.804	11.750	–	0.053600	1	0
2	150933		11.994	12.000	–	–	0	0
2	A151038	12.517	12.517		–	–	0	1
2	151038		12.900	13.017	–	–	0	0
2	A151022	13.250	13.250		–	–	0	1
2	151022		13.442	13.750	–	–	0	0

Case M08: optimal tour for a special set incl. 14 orders, $z = 0.074$

v	node	TE	tA	tL	rA2	rB	S	L
2	A150826	7.500	7.500		–	–	0	1
2	150826		7.887	8.000	–	–	0	0
2	A150869	8.000	8.000		–	–	1	0
2	A150890	8.000	8.025		0.025133	–	2	0
2	150890		8.322	8.500	–	–	1	0
2	150869		8.418	8.500	–	–	0	0
2	A150922	8.583	8.583		–	–	0	1
2	150922		8.828	9.083	–	–	0	0
2	A150920	9.000	9.015		0.014763	–	0	1
2	150920		9.322	9.500	–	–	0	0
2	A150939	9.500	9.500		–	–	1	0
2	150939		9.564	10.000	–	–	0	0
2	A150946	9.600	9.600		–	–	0	1
2	150946		9.807	10.100	–	–	0	0
2	A150965	9.983	9.983		–	–	0	1
2	150965		10.082	10.483	–	–	0	0
2	A150970	10.250	10.250		–	–	1	0
2	A150977	10.250	10.284		0.034267	–	1	1
2	150970		10.726	10.750	–	–	0	1
2	150977		10.750	10.750	–	–	0	0
2	A150749	11.000	11.000		–	–	1	0
2	150749		11.060	11.500	–	–	0	0
2	A150933	11.500	11.500		–	–	1	0
2	150933		12.000	12.000	–	–	0	0
2	A151038	12.517	12.517		–	–	0	1
2	151038		13.017	13.017	–	–	0	0
2	A151022	13.250	13.250		–	–	0	1
2	151022		13.442	13.750	–	–	0	0

B.2.2 Tabular Results for the Heuristic Methods

B.2.2.1 Input Data for a Whole Day - Offline Analysis

The input data for 18 September, 2001[1] contain 298 orders.

■ Information about orders:

```
298
    A18        18   GEB79-LS 70   GEB79-LS 70  13.00   0.00 10  0  0   6.00   5.00
    A1          1   GEB79-LS 70   GEB79-LS 70  13.00   0.00 10  0  0   8.00   3.00
    A12        12   GEB79-LS 70   GEB79-LS 70  12.50   0.00 10  0  0   2.00   4.00
    A8          8   GEB79-LS 70   GEB79-LS 70  12.50   0.00 10  0  0   9.00   3.00
    A5          5   GEB79-LS 70   GEB79-LS 70  12.50   0.00 10  0  0   8.00   9.00
    A16        16   GEB79-LS 70   GEB79-LS 70  12.00   0.00 10  0  0   8.00   5.00
    A11        11   GEB79-LS 70   GEB79-LS 70  12.00   0.00 10  0  0  23.00  15.00
    A2          2   GEB79-LS 70   GEB79-LS 70  12.00   0.00 10  0  0   7.00   3.00
    A10        10   GEB79-LS 70   GEB79-LS 70  11.50   0.00 10  0  0   6.00   8.00
A150837    150837      CT-BB 69      M2-01 73   0.25   0.00  4  0  0   3.00   0.00
A150839    150839      CT-BB 69      M2-01 73   0.25   0.00  4  0  0   4.00   4.00
A150841    150841       CM-A 69      KC-40 80   0.17   0.00  4  0  0   3.00   0.00
A150843    150843      HO-OP 74      M2-01 73   0.25   0.00  2  0  0   3.00   0.00
A150851    150851      FR-03 11      CT-BB 69   1.50   0.00  4  0  0   0.00   0.00
A150852    150852      CM-01 69      KC-40 80   1.72   0.00  4  0  0   3.00   2.00
A150859    150859      M3-08 71        U-A  5   3.20   0.00  1  0  0  10.00   5.00
A150860    150860        U-A  5      M3-08 71   4.25   0.00  1  0  0   8.00   8.00
A150862    150862      FR-03 11      CT-BB 69   5.50   0.00  4  0  0   0.00   0.00
A150855    150855      NN-03 55      M4-03 30   6.50   0.00  1  0  0  17.00   6.00
A150866    150866      NC-03 55       CT-A 36   6.05   0.00  4  0  0   1.00   6.00
A150868    150868      CK-09 68      CA-OP 69   7.28   7.33  5  0  2   0.00   0.00
A150878    150878      CK-09 68      CA-OP 69   7.28   7.33  5  0  2   0.00   0.00
A150829    150829      KK-03 75      M3-OP 80   7.25   0.00  1  0  2  12.00   4.00
A150864    150864      KK-04 75     RK-KSN 90   7.25   0.00  1  0  1   9.00  15.00
A150874    150874      M3-08 71       ZE-A 42   7.50   0.00  1  0  0   1.00   0.00
A150842    150842      RA-02 87       M5-A 60   7.75   0.00  1  0  0   4.00   5.00
A150752    150752      M5-03 60        U-A  5   7.50   0.00  2  0  0   4.00   5.00
A150826    150826      HO-02  5   RD-ANGIO 87   7.50   0.00  1  0  0   4.00   2.00
A150847    150847      NC-02 55       RK-A 85   7.50   0.00  1  0  0   2.00   9.00
A150873    150873      M1-01 80     NN-EMG 55   7.75   0.00  1  0  0  12.00   2.00
A150846    150846      NC-02 55       M5-A 60   8.00   0.00  1  0  0  12.00   4.00
A150886    150886      NN-02 55    M2-FUSS 86   8.00   0.00  1  0  0   2.00   7.00
A150887    150887      NN-02 55        U-A  5   8.00   0.00  2  0  0   3.00   5.00
A150857    150857      NN-03 55   M3-ANGIO 71   8.00   0.00  1  0  0   7.00   8.00
A150898    150898      M3-02 80    M5-LUFU 60   7.92   0.00  1  0  0   2.00   4.00
A150869    150869      M2-02 73    M5-ENDO 60   8.00   0.00  2  0  0   0.00   0.00
A150889    150889      M2-01 73       M2-A 86   8.00   0.00  1  0  0   6.00   6.00
A150870    150870      M2-02 73    M3-KARD 71   8.00   0.00  1  0  0   4.00   5.00
A150890    150890      FR-01 11       M2-A 86   8.00   0.00  2  0  0  14.00   1.00
A150894    150894       A-03  8      M4-03 81   8.00   0.00  2  0  0   0.00   4.00
A150910    150910    M3-KARD 71      M3-OP 80   8.22   0.00  2  0  0  14.00   7.00
A150902    150902      M3-06 80     M3-ANG 71   8.25   0.00  1  0  0   7.00   9.00
A150838    150838      RA-01 87       HO-A  5   8.25   0.00  2  0  0   4.00   0.00
A150899    150899      M4-02  5    M3-ECHO 80   7.95   0.00  2  0  0   0.00   0.00
A150909    150909      HO-05  5      NR-CT 55   8.25   0.00  1  0  0  15.00   5.00
A150908    150908      M2-03 73    M3-ECHO 80   8.25   0.00  1  0  0  13.00   4.00
```

---

[1]For the description of the content of the columns see Section B.B.1.1.

A150913	150913	M3-08 71	NN-ANGIO 55	8.35	0.00	1	0	0	6.00	37.00
A150900	150900	NP-06 55	M2-ESKOP 86	8.25	0.00	2	0	0	1.00	2.00
A150901	150901	NP-06 55	M2-ESKOP 86	8.25	0.00	2	0	0	0.00	6.00
A150912	150912	KK-04 75	CK-OP 69	8.33	0.00	1	0	1	8.00	6.00
A150865	150865	KK-04 75	RK-KSN 90	9.00	0.00	1	0	1	12.00	10.00
A150877	150877	D-04 21	HO-A 5	8.25	0.00	2	0	0	0.00	0.00
A150915	150915	M3-08 71	M3-ECHO 80	8.40	0.00	2	0	0	6.00	0.00
A150918	150918	M3-KARD 71	M3-OP 80	8.45	0.00	2	0	0	5.00	0.00
A150799	150799	FR-02 76	M5-ENDO 60	8.50	0.00	1	0	0	8.00	2.00
A150867	150867	FR-03 11	NN-A 55	8.50	0.00	2	0	0	4.00	0.00
A150921	150921	FR-01 11	RD-ROE57 69	8.57	0.00	2	0	0	0.00	0.00
A150916	150916	NN-03 55	M3-ECHO 80	8.43	0.00	2	0	0	1.00	0.00
A150917	150917	M5-03 60	M2-ESKOP 86	8.75	0.00	2	0	0	3.00	37.00
A150914	150914	M5-02 60	RD-CT49 87	8.50	0.00	2	0	0	7.00	0.00
A150748	150748	M5-03 60	HO-A 5	8.50	0.00	2	0	1	1.00	0.00
A150920	150920	KK-05 75	RK-KSN 90	9.00	0.00	1	0	1	15.00	5.00
A150853	150853	M4-02 5	M1-A 12	8.50	0.00	1	0	0	15.00	5.00
A150854	150854	M4-02 5	CU-A 69	8.50	0.00	1	0	0	7.00	9.00
A150919	150919	M5-03 60	M3-ANG 71	8.75	0.00	2	0	0	0.00	0.00
A150925	150925	RK-KSN 90	KK-04 75	8.77	0.00	1	0	1	12.00	4.00
A150926	150926	KK-04 11	RK-KSN 90	8.80	0.00	2	0	0	0.00	6.00
A150892	150892	M3-04 80	M5-A 60	8.75	0.00	1	0	0	5.00	6.00
A150893	150893	M3-04 80	M5-A 60	8.75	0.00	1	0	0	10.00	9.00
A150881	150881	M1-05 80	M5-A 60	8.75	0.00	2	0	0	0.00	0.00
A150880	150880	M1-05 80	M5-A 60	8.75	0.00	2	0	0	5.00	1.00
A150891	150891	M5-02 60	HO-A 5	9.00	0.00	1	0	0	3.00	6.00
A150751	150751	M5-03 60	M3-ANG 71	9.00	0.00	3	0	0	5.00	9.00
A150871	150871	M2-02 73	M5-LFA 60	8.50	0.00	1	0	0	5.00	4.00
A150750	150750	M5-03 60	RN-A 31	9.50	0.00	2	0	0	0.00	5.00
A150922	150922	RK-A 85	NC-02 55	8.58	0.00	1	0	0	12.00	8.00
A150930	150930	NN-02 55	CU-A 69	8.95	0.00	2	0	0	0.00	0.00
A150929	150929	A-02 8	CA-A 69	9.00	0.00	2	0	0	0.00	0.00
A150907	150907	CH-02 34	M4-02 5	8.75	0.00	1	0	0	8.00	7.00
A150844	150844	RA-02 87	HO-A 5	8.75	0.00	3	0	0	9.00	5.00
A150928	150928	M3-OP 80	KK-03 75	8.88	0.00	1	0	0	9.00	3.00
A150923	150923	M3-KARD 71	M3-ECHO 80	8.72	0.00	2	0	0	1.00	1.00
A150897	150897	FR-01 76	RN-A 31	9.50	0.00	1	0	0	2.00	4.00
A150934	150934	FR-01 76	M3-ANGIO 71	9.75	0.00	1	0	0	5.00	2.00
A150882	150882	FR-01 11	M2-A 86	9.00	0.00	2	0	0	0.00	0.00
A150883	150883	FR-01 11	RN-A 31	9.50	0.00	2	0	0	3.00	37.00
A150924	150924	M2-02 73	M2-ESKOP 86	8.83	0.00	1	0	0	9.00	6.00
A150885	150885	M2-03 73	HO-A 5	9.25	0.00	1	0	0	5.00	5.00
A150931	150931	M2-A 86	NN-02 55	9.05	0.00	1	0	0	16.00	4.00
A150856	150856	NN-03 55	RN-A 31	9.50	0.00	1	0	0	13.00	8.00
A150937	150937	NN-EMG 55	M1-01 80	9.50	0.00	1	0	0	12.00	3.00
A150938	150938	M3-ECHO 80	M2-03 73	9.30	0.00	1	0	0	13.00	5.00
A150936	150936	M4-01 30	FR-A 76	9.42	0.00	1	0	0	8.00	2.00
A150788	150788	U-04 5	RN-A 31	9.75	0.00	1	0	0	10.00	15.00
A150941	150941	M3-ANGIO 71	M5-03 60	9.42	0.00	2	0	0	1.00	0.00
A150944	150944	CM-01 69	M5-03 60	9.48	0.00	1	0	0	4.00	3.00
A150945	150945	M3-ECHO 80	M4-02 5	9.58	0.00	2	0	0	5.00	1.00
A150939	150939	M2-03 73	U-A 5	9.50	0.00	2	0	0	0.00	0.00
A150946	150946	M3-ANGIO 71	A-A 8	9.60	0.00	1	0	0	12.00	7.00
A150952	150952	HO-OP 74	M5-01 60	9.73	0.00	2	0	4	0.00	0.00
A150950	150950	M5-ENDO 60	RA-02 87	9.75	0.00	1	0	0	9.00	5.00
A150935	150935	CK-OP 69	CK-09 68	9.50	0.00	5	0	3	7.00	0.00
A150943	150943	D-04 21	RD-CT49 87	9.50	0.00	1	0	0	8.00	5.00

A150948	150948	M2-ESKOP	86	NP-06	55	9.67	0.00	1	0	0	13.00	7.00
A150947	150947	M2-ESKOP	86	M2-03	73	9.60	0.00	1	0	0	3.00	2.00
A150951	150951	HO-A	5	M5-03	60	9.75	0.00	3	0	0	3.00	6.00
A150961	150961	HO-A	5	D-04	21	10.00	0.00	2	0	0	0.00	1.00
A150955	150955	M5-LUFU	60	NC-02	55	9.80	0.00	1	0	0	10.00	6.00
A150958	150958	M5-LUFU	60	M3-02	80	9.82	0.00	1	0	0	5.00	9.00
A150969	150969	D-01	77	CA-A	69	10.25	0.00	3	0	0	7.00	6.00
A150967	150967	CU-A	69	M4-02	5	10.02	0.00	1	0	0	12.00	3.00
A150949	150949	NR-CT	55	HO-05	5	9.75	0.00	1	0	0	13.00	6.00
A150968	150968	M3-ECHO	80	NN-03	55	10.05	0.00	2	0	0	3.00	4.00
A150963	150963	NR-ANGIO	55	M3-08	71	10.00	0.00	1	0	0	5.00	4.00
A150960	150960	RK-KSN	90	KK-04	75	9.87	0.00	1	0	1	12.00	7.00
A150979	150979	KK-04	11	RK-KSN	90	10.38	0.00	2	0	0	5.00	0.00
A150959	150959	KK-04	75	RK-KSN	90	10.42	0.00	1	0	1	7.00	2.00
A150876	150876	O-05	63	RN-A	31	9.50	0.00	1	0	0	6.00	2.00
A150973	150973	RD-CT49	87	M5-02	60	10.18	0.00	2	0	0	5.00	0.00
A150975	150975	RN-A	31	FR-01	76	10.20	0.00	1	0	0	3.00	5.00
A150977	150977	M3-ANGIO	71	FR-01	76	10.25	0.00	1	0	0	16.00	8.00
A150953	150953	ZE-A	42	M3-08	71	9.77	0.00	2	0	0	0.00	0.00
A150964	150964	M3-ECHO	80	M3-08	71	9.98	0.00	2	0	0	7.00	1.00
A150965	150965	M3-ANGIO	71	M3-06	80	9.98	0.00	1	0	0	2.00	3.00
A150971	150971	M2-ESKOP	86	M2-03	73	10.15	0.00	1	0	0	5.00	2.00
A150980	150980	RD-ANGIO	87	HO-05	5	10.40	0.00	1	0	0	5.00	4.00
A150974	150974	M4-02	5	M3-ECHO	80	10.20	0.00	1	0	0	4.00	6.00
A150957	150957	CT-BB	69	NC-01	90	10.00	0.00	4	0	0	4.00	0.00
A150970	150970	CA-04	69	FR-A	11	10.25	0.00	2	0	0	0.00	0.00
A150984	150984	M2-A	86	FR-01	11	10.43	0.00	2	0	0	0.00	0.00
A150986	150986	RK-KSN	90	KK-05	75	10.62	0.00	1	0	2	2.00	7.00
A150954	150954	D-03	21	RD-A	87	10.00	0.00	1	0	0	2.00	4.00
A150987	150987	RN-A	31	U-04	5	10.60	0.00	1	0	0	2.00	5.00
A150990	150990	M4-02	5	RD-ANGIO	87	10.75	0.00	1	0	0	4.00	5.00
A150985	150985	M3-ANGIO	71	M5-03	60	10.52	0.00	3	0	0	13.00	4.00
A150983	150983	M3-ANGIO	71	CA-A	69	10.43	0.00	1	0	0	13.00	6.00
A150749	150749	M5-03	60	RA-A	87	11.00	0.00	2	0	0	0.00	0.00
A150888	150888	M5-02	60	RA-A	87	11.00	0.00	1	0	0	8.00	6.00
A150998	150998	M5-LUFU	60	M1-05	80	11.07	0.00	2	0	0	5.00	0.00
A150966	150966	RA-02	87	D-A	78	10.25	0.00	3	0	0	7.00	4.00
A150989	150989	RN-A	31	NN-03	55	10.75	0.00	1	0	0	8.00	13.00
A150988	150988	NN-04	55	RA-PA	87	10.75	0.00	3	0	0	8.00	5.00
A150996	150996	NN-A	55	FR-03	11	11.00	0.00	2	0	1	14.00	3.00
A150993	150993	U-A	5	NN-02	55	10.95	0.00	2	0	0	3.00	4.00
A150992	150992	NP-04	55	M3-KARD	71	11.00	0.00	1	0	1	7.00	3.00
A150991	150991	M5-ENDO	60	FR-02	76	10.88	0.00	1	0	0	5.00	4.00
A150845	150845	RA-02	87	FR-A	76	11.00	0.00	1	0	0	10.00	2.00
A150999	150999	RD-CT57	69	D-01	77	11.12	0.00	3	0	0	23.00	15.00
A151001	151001	RD-CT49	87	D-04	21	11.17	0.00	1	0	0	4.00	7.00
A150927	150927	O-01	63	M2-A	86	11.00	0.00	1	0	0	12.00	3.00
A150997	150997	U-A	5	M2-03	73	11.05	0.00	2	0	0	0.00	0.00
A150978	150978	CK-OP	69	CK-09	68	10.75	0.00	5	0	2	0.00	0.00
A150850	150850	KK-07	75	RK-KSN	90	11.25	0.00	1	0	1	3.00	4.00
A151010	151010	KK-04	11	RK-KSN	90	11.48	0.00	2	0	0	1.00	0.00
A151005	151005	RK-KSN	90	KK-04	75	11.28	0.00	1	0	1	3.00	4.00
A151006	151006	M5-01	60	HO-OP	74	11.28	0.00	2	0	4	5.00	1.00
A151003	151003	RN-A	31	O-05	63	11.25	0.00	1	0	0	12.00	5.00
A150994	150994	HO-03	5	RD-DL	87	11.20	11.25	1	0	0	12.00	6.00
A151017	151017	M3-KARD	71	M3-ECHO	80	11.72	0.00	-1	0	0	8.00	3.00
A150995	150995	M3-KARD	71	M3-ECHO	80	10.98	0.00	1	0	0	9.00	5.00

A151008	151008	FR-01	11	RK-A	85	11.45	0.00	2	0	0	0.00	0.00
A151012	151012	FR-A	76	M4-01	30	11.57	0.00	1	0	0	7.00	7.00
A151000	151000	M2-ESKOP	86	M2-02	73	11.25	0.00	1	0	0	9.00	9.00
A151002	151002	M2-ESKOP	86	M2-03	73	11.17	0.00	1	0	0	13.00	6.00
A151015	151015	D-03	21	FR-OP	76	11.67	0.00	1	0	0	12.00	7.00
A151009	151009	U-04	5	RD-ROE49	86	11.50	0.00	1	0	0	2.00	4.00
A150911	150911	M4-03	30	NN-03	55	11.50	0.00	1	0	0	7.00	3.00
A150840	150840	RA-01	87	NN-A	55	11.50	0.00	2	0	0	0.00	0.00
A151014	151014	M5-LUFU	60	M3-04	80	11.65	0.00	1	0	0	9.00	7.00
A151023	151023	A-A	8	NN-03	55	11.82	0.00	1	0	0	3.00	4.00
A150972	150972	NN-02	55	RK-A	85	11.50	0.00	1	0	0	7.00	7.00
A150933	150933	O-05	63	RD-ANGIO	87	11.50	0.00	2	0	0	14.00	3.00
A151007	151007	NC-01	90	M5-LUFU	60	11.35	0.00	1	0	1	8.00	4.00
A151028	151028	NC-01	90	GK-OP	89	12.25	0.00	-1	0	1	5.00	7.00
A151029	151029	HO-02	5	NN-A	55	12.50	0.00	-1	0	0	11.00	5.00
A151016	151016	RD-ROE49	86	D-03	21	11.70	0.00	1	0	0	5.00	5.00
A151018	151018	M5-02	60	M3-OP	80	11.75	0.00	1	0	0	2.00	4.00
A151019	151019	M5-02	60	M3-OP	80	11.75	0.00	1	0	0	3.00	4.00
A151020	151020	M3-OP	80	M3-08	71	12.00	0.00	1	0	0	6.00	1.00
A151034	151034	D-A	78	RA-02	87	12.40	0.00	3	0	0	12.00	7.00
A151027	151027	M5-LUFU	60	M3-04	80	12.25	0.00	1	0	0	10.00	8.00
A151011	151011	M2-A	86	M2-01	73	11.48	0.00	1	0	0	3.00	4.00
A151030	151030	M2-03	73	M3-A	71	12.50	0.00	1	0	0	8.00	6.00
A151043	151043	KK-07	11	RK-KSN	90	12.38	0.00	2	0	0	4.00	0.00
A151032	151032	RK-KSN	90	KK-07	75	12.38	0.00	1	0	0	12.00	6.00
A151024	151024	KK-05	75	RK-KSN	90	12.50	0.00	1	0	1	11.00	4.00
A150879	150879	NN-04	55	FR-PA	11	11.75	0.00	2	0	0	10.00	0.00
A151036	151036	FR-02	76	RD-CT57	69	12.48	0.00	1	0	0	13.00	8.00
A151004	151004	FR-01	76	RN-A	31	12.50	0.00	1	0	0	13.00	5.00
A151042	151042	RD-DL	87	FR-01	11	12.60	0.00	2	0	0	2.00	6.00
A151021	151021	GEB49-RB	87	M5-02	60	12.00	0.00	1	0	0	3.00	4.00
A151033	151033	M5-LUFU	60	M2-02	73	12.50	0.00	1	0	0	12.00	1.00
A151031	151031	M2-A	86	O-02	63	12.37	0.00	1	0	0	2.00	6.00
A151040	151040	M3-ECHO	80	M4-02	5	12.58	0.00	1	0	0	15.00	2.00
A151035	151035	HO-A	5	M2-03	73	12.48	0.00	1	0	0	3.00	7.00
A151038	151038	M2-A	86	O-02	63	12.52	0.00	1	0	0	18.00	6.00
A150895	150895	U-01	5	RN-A	31	12.75	0.00	1	0	0	8.00	6.00
A151041	151041	HO-A	5	RA-02	87	12.75	0.00	3	0	0	2.00	3.00
A151044	151044	RD-DL	87	HO-03	5	12.78	0.00	1	0	0	5.00	3.00
A151037	151037	A-03	8	U-A	5	12.75	0.00	2	0	0	3.00	0.00
A151054	151054	RD-ANGIO	87	M4-02	5	13.13	0.00	1	0	0	9.00	4.00
A151046	151046	M3-ECHO	80	M3-KARD	71	12.93	0.00	1	0	0	8.00	7.00
A151048	151048	CA-A	69	A-02	8	13.00	0.00	2	0	0	3.00	0.00
A151051	151051	CA-A	69	M2-02	73	13.07	0.00	1	0	0	15.00	5.00
A151050	151050	M2-02	73	M2-E	86	13.07	0.00	1	0	0	2.00	5.00
A151056	151056	M3-KARD	71	NP-04	55	13.20	0.00	1	0	0	3.00	15.00
A151039	151039	NN-05	55	M4-03	30	12.75	0.00	1	0	0	3.00	3.00
A150982	150982	M4-03	30	A-03	8	12.67	0.00	1	0	0	5.00	10.00
A151013	151013	U-04	5	RN-A	31	13.25	0.00	1	0	0	8.00	8.00
A151058	151058	RD-CT49	87	U-04	5	13.30	0.00	1	0	0	5.00	4.00
A151060	151060	NN-A	55	HO-02	5	13.50	0.00	1	0	3	9.00	9.00
A151055	151055	D-03	21	M5-A	60	13.25	0.00	1	0	0	2.00	7.00
A151059	151059	FR-PA	11	NN-04	55	13.43	0.00	2	0	0	6.00	0.00
A151049	151049	RD-ANGIO	87	O-05	63	13.05	0.00	1	0	0	8.00	4.00
A151026	151026	O-05	63	RN-A	31	13.25	0.00	1	0	0	6.00	7.00
A150906	150906	O-02	63	RK-A	85	13.75	0.00	1	0	0	7.00	10.00
A151045	151045	CU-01	69	U-A	5	13.08	0.00	1	0	0	8.00	6.00

A151047	151047	M1-A	12	M4-02	5	13.17	0.00	1	0	0	8.00 3.00
A151053	151053	RK-A	85	NN-04	55	13.10	0.00	3	0	0	7.00 5.00
A151052	151052	M2-ESKOP	86	M2-02	73	13.25	0.00	1	0	0	2.00 6.00
A151061	151061	D-04	21	NN-A	55	13.75	0.00	1	0	0	5.00 9.00
A151064	151064	RK-A	85	NN-02	55	13.80	0.00	1	0	0	12.00 5.00
A151022	151022	NN-03	55	RN-PET	31	13.25	0.00	1	0	0	9.00 6.00
A150858	150858	NN-03	55	M4-A	30	14.00	0.00	1	0	0	5.00 8.00
A151063	151063	RK-A	85	FR-01	11	14.00	0.00	2	0	0	1.00 5.00
A151066	151066	RA-A	87	M5-03	60	14.00	0.00	2	0	0	3.00 1.00
A151068	151068	D-03	21	RD-A	87	14.00	0.00	1	0	0	12.00 7.00
A151065	151065	RD-CT57	69	FR-02	76	13.85	0.00	1	0	0	2.00 15.00
A151071	151071	U-A	5	A-03	8	14.13	0.00	2	0	0	3.00 0.00
A151067	151067	M3-ANGIO	71	M2-03	73	14.00	0.00	1	0	0	4.00 5.00
A151072	151072	RD-FRKI	11	M5-02	60	14.28	0.00	2	0	0	4.00 0.00
A151069	151069	KK-04	11	CK-OP	69	14.25	0.00	2	0	0	3.00 0.00
A151070	151070	CK-OP	69	KK-04	75	14.50	0.00	1	0	2	3.00 4.00
A151073	151073	RD-FRKI	76	RA-02	87	14.37	0.00	1	0	0	5.00 8.00
A151076	151076	FR-02	76	GEB40-EC	80	14.43	0.00	1	0	0	8.00 6.00
A151077	151077	RN-A	31	RD-ROE57	69	14.43	0.00	1	0	0	17.00 7.00
A151079	151079	RD-ROE49	86	D-03	21	14.48	0.00	1	0	0	15.00 4.00
A151078	151078	M3-KARD	71	D-04	21	14.45	0.00	1	0	0	13.00 4.00
A151074	151074	M2-ESKOP	86	M2-02	73	14.38	0.00	1	0	0	3.00 5.00
A151080	151080	M2-ESKOP	86	M2-03	73	14.48	0.00	1	0	0	9.00 4.00
A151082	151082	M3-08	71	RK-KSN	90	14.75	0.00	2	0	0	0.00 2.00
A151081	151081	RK-KSN	90	KK-05	75	14.55	0.00	1	0	1	5.00 7.00
A151083	151083	KK-05	11	RK-KSN	90	14.62	0.00	2	0	0	14.00 0.00
A151084	151084	RN-A	31	FR-01	76	14.67	0.00	1	0	0	9.00 5.00
A151089	151089	RD-FRKI	11	M5-02	60	15.02	0.00	2	0	0	4.00 0.00
A151087	151087	M3-OP	80	M5-02	60	15.00	0.00	1	0	0	5.00 4.00
A151085	151085	RN-A	31	O-05	63	15.00	0.00	1	0	0	5.00 4.00
A150848	150848	O-02	63	RK-A	85	14.75	0.00	1	0	0	12.00 10.00
A151090	151090	RD-CT57	69	U-04	5	15.25	0.00	1	0	1	12.00 4.00
A151092	151092	NN-A	55	D-04	21	15.25	0.00	1	0	0	2.00 6.00
A151093	151093	M3-ECHO	80	FR-02	76	15.25	0.00	1	0	0	2.00 3.00
A151091	151091	RK-A	85	O-02	63	15.17	0.00	1	0	0	7.00 5.00
A151096	151096	RN-A	31	NN-03	55	15.37	0.00	1	0	0	6.00 4.00
A151086	151086	NN-02	55	RK-A	85	15.50	0.00	1	0	0	7.00 15.00
A151094	151094	NN-03	55	RK-A	85	15.50	0.00	3	0	0	8.00 4.00
A151099	151099	CM-A	69	M5-03	60	15.40	0.00	1	0	0	10.00 5.00
A151095	151095	CT-BB	69	M5-02	60	15.50	0.00	4	0	0	5.00 4.00
A151097	151097	M5-ENDO	60	D-03	21	15.67	0.00	1	0	0	17.00 6.00
A151102	151102	M2-ESKOP	86	M2-02	73	15.67	0.00	1	0	0	9.00 8.00
A151101	151101	A-01	8	RD-ROE49	86	15.50	0.00	1	0	0	6.00 6.00
A151104	151104	M4-A	30	NN-03	55	15.83	0.00	1	0	0	7.00 6.00
A151106	151106	NC-02	55	RK-A	85	15.98	0.00	2	0	0	0.00 0.00
A151107	151107	RN-A	31	U-01	5	16.00	0.00	1	0	0	7.00 11.00
A151103	151103	M2-01	73	RD-CT57	69	16.00	0.00	1	0	0	15.00 10.00
A151108	151108	CM-A	69	M1-01	80	16.07	0.00	1	0	0	5.00 5.00
A151110	151110	RD-ROE49	86	A-01	8	16.23	0.00	1	0	0	8.00 8.00
A151116	151116	U-A	5	CU-01	69	16.72	0.00	1	0	0	7.00 5.00
A151115	151115	M3-OP	80	M5-02	60	16.75	0.00	1	0	0	7.00 4.00
A151112	151112	RK-KS	85	O-02	63	16.58	0.00	1	0	0	8.00 8.00
A151117	151117	RD-CT57	69	M2-01	73	17.00	0.00	1	0	0	9.00 5.00
A151114	151114	CT-BB	69	U-LABOR	5	16.75	0.00	4	0	0	0.00 4.00
A151119	151119	NC-03	55	RK-A	85	17.50	0.00	1	0	0	6.00 6.00
A151118	151118	KK-04	75	NR-CT	55	17.50	0.00	1	0	2	6.00 4.00
A151120	151120	RK-A	85	NN-02	55	17.65	0.00	1	0	1	4.00 6.00

```
A151121  151121   RK-KSN 90   M3-08 71 17.75  0.00  2  0  0  0.00  0.00
A151122  151122   M4-03  30   NN-05 55 18.00  0.00  1  0  0  9.00  8.00
A151125  151125    RK-A  85   NN-03 55 18.02  0.00  2  0  1  3.00  0.00
A151127  151127    RK-A  85   NC-02 55 18.40  0.00  2  0  0  2.00  0.00
A151123  151123 M2-ESKOP 86   FR-05 76 17.83  0.00  1  0  0  9.00  2.00
A151128  151128   O-06  63    RK-A  85 18.50  0.00  1  0  0  3.00  9.00
A151129  151129   CH-01 69    KK-01 11 18.52  0.00  4  0  0  3.00  0.00
A151131  151131   NR-CT 55    KK-04 75 19.00  0.00  1  0  0  7.00  6.00
A151130  151130   M3-04 80    RK-KSN 90 19.00  0.00 1  0  0  5.00 13.00
A151132  151132    RK-A 85    NC-03 55 19.05  0.00  2  0  1  7.00  1.00
A151133  151133   A-03   8    RD-CT57 69 19.53 0.00  2  0  0  0.00  0.00
A151135  151135   CT-BB 69    KK-01 11 19.87  0.00 -4  0  0  1.00  0.00
A151136  151136   CT-BB 69    M5-03 60 20.25  0.00  4  0  0  0.00  0.00
A151137  151137 RD-ROE57 69   A-03   8 20.38  0.00  2  0  0  0.00  0.00
A151138  151138    RK-A 85    O-06  63 20.73  0.00  1  0  0 10.00  5.00
A151139  151139   RK-KSN 90   M3-04 80 20.72  0.00  1  0  0  9.00  6.00
A151144  151144    CU-A 69    NN-A  55 21.93  0.00  2  0  1  3.00  6.00
A151534  151534    CM-A 69    KC-40 80  5.88  0.00  4  0  0  0.00  0.00
A151146  151146   CU-OP 69    CK-09 68 22.42  0.00  5  0  1  0.00  0.00
A151148  151148   NC-01 90    CT-BB 69 23.00  0.00 -4  0  0  1.00  0.00
A151145  151145   CT-BB 69    M1-03 80 22.32  0.00  4  0  0  3.00  0.00
A151149  151149   M2-01 73    KC-40 80 23.00  0.00  4  0  0  2.00  2.00
A151147  151147   CM-01 69    KC-40 80 22.68  0.00  4  0  0  6.00  0.00
A151150  151150    NN-A 55    CM-01 69 23.25  0.00  2  0  0  1.00  0.00
```

- Information about vehicles include the number of vehicles (11 vehicles), the current time (0), the time when the vehicles are available, and their capacities for lying, sitting and wheelchair passengers:

```
11   0
1    70    0        75600   1   3   0
2    70    25200    51120   1   3   0
3    70    25200    86340   1   3   0
4    70    27000    56520   1   5   1
5    70    28080    57600   1   5   1
6    70    28080    57600   1   5   1
7    70    28800    58320   1   5   1
8    70    28800    58320   1   5   1
9    70    28800    58320   1   5   1
10   70    30600    60120   1   5   1
11   70    30600    60120   1   5   1
```

B.2.2.2   Results for CIH and SA

Here we display the final tours for some combinations of CIH and SA, applied to the input data of 18 September, 2001 (see page 144).

The table of results obtained when SA applied after the CIH:

v	s-node	d-node	tD	TD	tA	TT	delay	z	S	L	IC
1	DS1	A150841	0.140	0.026	0.167	0.167	–	0.000	0	0	0
1	A150841	A150837	0.217	0.000	0.250	0.250	–	0.000	0	0	0
1	A150837	A150839	0.300	0.000	0.300	0.250	0.050	0.050	0	0	0
1	A150839	A150843	0.367	0.060	0.427	0.250	0.177	0.227	1	0	0
1	A150843	150839	0.477	0.039	0.516	0.750	–	0.227	1	0	13

1	150839	150843	0.583	0.000	0.583	0.750	–	0.227	0 0 10
1	150843	150841	0.583	0.014	0.596	0.667	–	0.227	0 0 26
1	150841	150837	0.596	0.073	0.670	0.750	–	0.227	0 0 26
1	150837	A150851	0.670	0.025	1.500	1.500	–	0.227	0 0 0
1	A150851	A150852	1.500	0.058	1.717	1.717	–	0.227	0 0 0
1	A150852	150852	1.767	0.023	1.790	2.217	–	0.227	0 0 5
1	150852	150851	1.823	0.057	1.881	2.000	–	0.227	0 0 23
1	150851	A150859	1.881	0.034	3.200	3.200	–	0.227	0 1 0
1	A150859	150859	3.367	0.028	3.395	3.700	–	0.227	0 0 12
1	150859	A150860	3.478	0.000	4.250	4.250	–	0.227	0 1 0
1	A150860	150860	4.383	0.036	4.419	4.750	–	0.227	0 0 11
1	150860	A150862	4.553	0.060	5.500	5.500	–	0.227	0 0 0
1	A150862	A151534	5.500	0.058	5.883	5.883	–	0.227	0 0 0
1	A151534	150862	5.883	0.000	5.883	6.000	–	0.227	0 0 23
1	150862	A150866	5.883	0.019	6.050	6.050	–	0.227	0 0 0
1	A150866	151534	6.067	0.040	6.107	6.383	–	0.227	0 0 14
1	151534	150866	6.107	0.044	6.151	6.550	–	0.227	0 0 7
1	150866	A150855	6.251	0.032	6.500	6.500	–	0.227	0 1 0
1	A150855	150855	6.783	0.047	6.830	7.000	–	0.227	0 0 20
1	150855	A150864	6.930	0.065	7.250	7.250	–	0.227	1 1 0
1	A150864	A150874	7.400	0.031	7.500	7.500	–	0.227	2 1 0
1	A150874	150864	7.517	0.085	7.602	7.750	–	0.227	1 0 22
1	150864	150874	7.852	0.032	7.884	8.000	–	0.227	0 0 24
1	150874	A150899	7.884	0.055	7.950	7.950	–	0.227	1 0 0
1	A150899	A150889	7.950	0.042	8.000	8.000	–	0.227	1 1 0
1	A150889	150899	8.100	0.014	8.114	8.450	–	0.227	0 1 10
1	150899	150889	8.114	0.063	8.177	8.500	–	0.227	0 0 11
1	150889	A150912	8.277	0.068	8.345	8.333	0.012	0.239	1 1 0
1	A150912	A150867	8.478	0.017	8.500	8.500	–	0.239	2 1 0
1	A150867	150912	8.567	0.058	8.625	8.833	–	0.239	1 0 17
1	150912	150867	8.725	0.019	8.743	9.000	–	0.239	0 0 15
1	150867	A150925	8.743	0.011	8.767	8.767	–	0.239	1 1 0
1	A150925	150925	8.967	0.057	9.023	9.267	–	0.239	0 0 16
1	150925	A150931	9.090	0.055	9.145	9.050	0.095	0.334	0 1 0
1	A150931	150931	9.412	0.081	9.492	9.550	–	0.334	0 0 21
1	150931	A150936	9.559	0.047	9.606	9.417	0.189	0.523	0 1 0
1	A150936	150936	9.739	0.025	9.764	9.917	–	0.523	0 0 10
1	150936	A150788	9.797	0.037	9.834	9.750	0.084	0.607	0 1 0
1	A150788	A150964	10.001	0.041	10.042	9.983	0.059	0.667	1 1 0
1	A150964	150788	10.159	0.061	10.220	10.250	–	0.667	1 0 24
1	150788	150964	10.470	0.017	10.487	10.483	0.004	0.670	0 0 27
1	150964	A150959	10.504	0.053	10.557	10.417	0.141	0.811	1 1 0
1	A150959	150959	10.674	0.057	10.731	10.917	–	0.811	0 0 11
1	150959	A150989	10.764	0.046	10.810	10.750	0.060	0.871	0 1 0
1	A150989	150989	10.943	0.078	11.022	11.250	–	0.871	0 0 13
1	150989	A151000	11.238	0.044	11.282	11.250	0.032	0.903	0 1 0
1	A151000	151000	11.432	0.078	11.510	11.750	–	0.903	0 0 14
1	151000	A151017	11.660	0.029	11.717	11.717	–	0.903	0 1 0
1	A151017	151017	11.850	0.065	11.915	12.217	–	0.903	0 0 12
1	151017	A151020	11.965	0.000	12.000	12.000	–	0.903	0 1 0
1	A151020	151020	12.100	0.015	12.115	12.500	–	0.903	0 0 7
1	151020	A2	12.132	0.039	12.171	12.000	0.171	1.074	3 1 0
1	A2	2	12.287	0.000	12.287	12.500	–	1.074	0 0 8
1	2	A151030	12.337	0.033	12.500	12.500	–	1.074	0 1 0
1	A151030	151030	12.633	0.029	12.662	13.000	–	1.074	0 0 10
1	151030	A150895	12.762	0.028	12.790	12.750	0.040	1.114	0 1 0
1	A150895	150895	12.923	0.072	12.995	13.250	–	1.114	0 0 13

1	150895	A151049	13.095	0.004	13.099	13.050	0.049	1.163	0 1	0
1	A151049	151049	13.233	0.085	13.318	13.550	–	1.163	0 0	14
1	151049	A151058	13.385	0.061	13.446	13.300	0.146	1.310	0 1	0
1	A151058	A151059	13.530	0.077	13.606	13.433	0.173	1.483	1 1	0
1	A151059	151058	13.706	0.039	13.746	13.800	–	1.483	1 0	18
1	151058	151059	13.812	0.065	13.878	13.933	–	1.483	0 0	17
1	151059	A150906	13.878	0.043	13.921	13.750	0.171	1.653	0 1	0
1	A150906	150906	14.037	0.054	14.092	14.250	–	1.653	0 0	11
1	150906	A151073	14.258	0.037	14.367	14.367	–	1.653	0 1	0
1	A151073	151073	14.450	0.072	14.522	14.867	–	1.653	0 0	10
1	151073	A151079	14.655	0.002	14.657	14.483	0.174	1.827	0 1	0
1	A151079	151079	14.907	0.075	14.982	14.983	–	1.827	0 0	20
1	151079	A151085	15.049	0.042	15.091	15.000	0.091	1.918	0 1	0
1	A151085	151085	15.175	0.081	15.256	15.500	–	1.918	0 0	10
1	151085	A151090	15.323	0.054	15.377	15.250	0.127	2.045	1 1	0
1	A151090	151090	15.577	0.057	15.634	15.750	–	2.045	0 0	16
1	151090	A151097	15.701	0.088	15.789	15.667	0.122	2.167	0 1	0
1	A151097	151097	16.072	0.077	16.148	16.167	–	2.167	0 0	22
1	151097	A151114	16.248	0.025	16.750	16.750	–	2.167	0 0	0
1	A151114	151114	16.750	0.057	16.807	17.250	–	2.167	0 0	4
1	151114	A151118	16.874	0.029	17.500	17.500	–	2.167	2 1	0
1	A151118	151118	17.600	0.046	17.646	18.000	–	2.167	0 0	9
1	151118	A151120	17.713	0.039	17.752	17.650	0.102	2.269	1 1	0
1	A151120	151120	17.818	0.045	17.864	18.150	–	2.269	0 0	7
1	151120	A151122	17.964	0.047	18.011	18.000	0.011	2.279	0 1	0
1	A151122	151122	18.161	0.077	18.237	18.500	–	2.279	0 0	14
1	151122	A151129	18.371	0.025	18.517	18.517	–	2.279	0 0	0
1	A151129	151129	18.567	0.035	18.601	19.017	–	2.279	0 0	6
1	151129	A151130	18.601	0.076	19.000	19.000	–	2.279	0 1	0
1	A151130	A151132	19.083	0.057	19.141	19.050	0.091	2.370	2 1	0
1	A151132	151132	19.257	0.045	19.303	19.550	–	2.370	0 1	10
1	151132	151130	19.319	0.011	19.330	19.500	–	2.370	0 0	20
1	151130	A151139	19.547	0.000	20.717	20.717	–	2.370	0 1	0
1	A151139	151139	20.867	0.044	20.910	21.217	–	2.370	0 0	12
1	151139	DD1	21.010	0.072	21.082	24.000	–	2.370	0 0	12
2	DS1	A150878	7.177	0.107	7.283	7.283	–	0.000	3 1	0
2	A150878	150878	7.283	0.003	7.287	7.333	–	0.000	0 0	1
2	150878	A150868	7.396	0.107	7.503	7.283	0.220	0.220	3 1	0
2	A150868	150868	7.503	0.003	7.506	7.333	0.173	0.392	0 0	1
2	150868	A150826	7.616	0.039	7.655	7.500	0.155	0.547	0 1	0
2	A150826	150826	7.722	0.076	7.798	8.000	–	0.547	0 0	9
2	150826	A150869	7.831	0.080	8.000	8.000	–	0.547	1 0	0
2	A150869	A150894	8.000	0.031	8.031	8.000	0.031	0.579	2 0	0
2	A150894	150869	8.031	0.074	8.105	8.500	–	0.579	1 0	7
2	150869	150894	8.105	0.079	8.184	8.500	–	0.579	0 0	10
2	150894	A150909	8.251	0.034	8.285	8.250	0.035	0.613	0 1	0
2	A150909	150909	8.535	0.065	8.600	8.750	–	0.613	0 0	19
2	150909	A150922	8.683	0.039	8.722	8.583	0.139	0.752	0 1	0
2	A150922	150922	8.922	0.045	8.967	9.083	–	0.752	0 0	15
2	150922	A150920	9.100	0.053	9.153	9.000	0.153	0.905	1 1	0
2	A150920	150920	9.403	0.057	9.460	9.500	–	0.905	0 0	19
2	150920	A150939	9.544	0.037	9.580	9.500	0.080	0.985	1 0	0
2	A150939	A150946	9.580	0.029	9.609	9.600	0.009	0.994	1 1	0
2	A150946	150939	9.809	0.028	9.837	10.000	–	0.994	0 1	16
2	150939	150946	9.837	0.038	9.876	10.100	–	0.994	0 0	16
2	150946	A150965	9.992	0.005	9.997	9.983	0.014	1.009	0 1	0

2	A150965	A150970	10.031	0.056	10.250	10.250	-	1.009	1 1	0
2	A150970	150970	10.250	0.035	10.285	10.750	-	1.009	0 1	3
2	150970	150965	10.285	0.076	10.361	10.483	-	1.009	0 0	22
2	150965	A150978	10.411	0.181	10.750	10.750	-	1.009	3 1	0
2	A150978	150978	10.750	0.003	10.753	11.250	-	1.009	0 0	1
2	150978	A150749	10.860	0.059	11.000	11.000	-	1.009	1 0	0
2	A150749	A150845	11.000	0.060	11.060	11.000	0.060	1.069	1 1	0
2	A150845	150749	11.227	0.000	11.227	11.500	-	1.069	0 1	14
2	150749	150845	11.227	0.030	11.257	11.500	-	1.069	0 0	12
2	150845	A151002	11.290	0.070	11.360	11.167	0.194	1.262	0 1	0
2	A151002	151002	11.577	0.078	11.655	11.667	-	1.262	0 0	18
2	151002	A151015	11.755	0.016	11.771	11.667	0.104	1.366	0 1	0
2	A151015	151015	11.971	0.036	12.007	12.167	-	1.366	0 0	15
2	151015	A151004	12.123	0.000	12.500	12.500	-	1.366	0 1	0
2	A151004	151004	12.717	0.068	12.785	13.000	-	1.366	0 0	18
2	151004	A151044	12.868	0.004	12.872	12.783	0.089	1.455	0 1	0
2	A151044	151044	12.955	0.045	13.000	13.283	-	1.455	0 0	8
2	151044	A151046	13.050	0.041	13.092	12.933	0.158	1.613	0 1	0
2	A151046	151046	13.225	0.015	13.240	13.433	-	1.613	0 0	9
2	151046	A151026	13.357	0.069	13.425	13.250	0.175	1.789	0 1	0
2	A151026	151026	13.525	0.057	13.583	13.750	-	1.789	0 0	10
2	151026	A151061	13.699	0.073	13.773	13.750	0.023	1.811	0 1	0
2	A151061	151061	13.856	0.036	13.892	14.250	-	1.811	0 0	8
2	151061	A151068	14.042	0.042	14.084	14.000	0.084	1.895	0 1	0
2	A151068	151068	14.284	0.046	14.330	14.500	-	1.895	0 0	15
2	151068	DD1	14.447	0.078	14.525	24.000	-	1.895	0 0	15
3	DS1	A150829	7.232	0.018	7.250	7.250	-	0.000	2 1	0
3	A150829	A150752	7.450	0.069	7.519	7.500	0.019	0.019	3 1	0
3	A150752	150829	7.586	0.074	7.660	7.750	-	0.019	1 0	25
3	150829	A150873	7.727	0.000	7.750	7.750	-	0.019	1 1	0
3	A150873	150752	7.950	0.039	7.989	8.000	-	0.019	0 1	29
3	150752	150873	8.072	0.065	8.137	8.250	-	0.019	0 0	24
3	150873	A150748	8.171	0.040	8.500	8.500	-	0.019	2 0	0
3	A150748	A150914	8.517	0.000	8.517	8.500	0.017	0.036	3 0	0
3	A150914	150914	8.633	0.060	8.693	9.000	-	0.036	2 0	11
3	150914	A150907	8.693	0.060	8.754	8.750	0.004	0.039	2 1	0
3	A150907	150748	8.887	0.048	8.935	9.000	-	0.039	0 1	27
3	150748	150907	8.935	0.000	8.935	9.250	-	0.039	0 0	11
3	150907	A150865	9.051	0.029	9.081	9.000	0.081	0.120	1 1	0
3	A150865	150865	9.281	0.057	9.338	9.500	-	0.120	0 0	16
3	150865	A150856	9.505	0.021	9.526	9.500	0.026	0.146	0 1	0
3	A150856	150856	9.742	0.042	9.784	10.000	-	0.146	0 0	16
3	150856	A150967	9.918	0.060	10.017	10.017	-	0.146	0 1	0
3	A150967	150967	10.217	0.057	10.274	10.517	-	0.146	0 0	16
3	150967	A150983	10.324	0.036	10.433	10.433	-	0.146	0 1	0
3	A150983	150983	10.650	0.056	10.706	10.933	-	0.146	0 0	17
3	150983	A150888	10.806	0.041	11.000	11.000	-	0.146	0 1	0
3	A150888	150888	11.133	0.060	11.193	11.500	-	0.146	0 0	12
3	150888	A151003	11.293	0.004	11.297	11.250	0.047	0.193	0 1	0
3	A151003	A150933	11.497	0.081	11.579	11.500	0.079	0.272	1 1	0
3	A150933	151003	11.812	0.000	11.812	11.750	0.062	0.334	1 0	31
3	151003	150933	11.895	0.061	11.957	12.000	-	0.334	0 0	23
3	150933	A151028	12.007	0.093	12.250	12.250	-	0.334	1 1	0
3	A151028	151028	12.333	0.030	12.364	12.750	-	0.334	0 0	7
3	151028	A151029	12.480	0.057	12.537	12.500	0.037	0.371	0 1	0
3	A151029	151029	12.720	0.065	12.785	13.000	-	0.371	0 0	15

3	151029	A18	12.869	0.042	13.000	13.000	–	0.371	3 1	0
3	A18	18	13.100	0.000	13.100	13.500	–	0.371	0 0	6
3	18	A151054	13.183	0.024	13.208	13.133	0.074	0.445	0 1	0
3	A151054	151054	13.358	0.045	13.403	13.633	–	0.445	0 0	12
3	151054	A150858	13.469	0.065	14.000	14.000	–	0.445	0 1	0
3	A150858	150858	14.083	0.047	14.130	14.500	–	0.445	0 0	8
3	150858	A151070	14.264	0.058	14.500	14.500	–	0.445	2 1	0
3	A151070	151070	14.550	0.036	14.586	15.000	–	0.445	0 0	6
3	151070	A151087	14.653	0.023	15.000	15.000	–	0.445	0 1	0
3	A151087	151087	15.083	0.075	15.158	15.500	–	0.445	0 0	10
3	151087	A151092	15.225	0.067	15.292	15.250	0.042	0.487	0 1	0
3	A151092	151092	15.325	0.042	15.367	15.750	–	0.487	0 0	5
3	151092	A151086	15.467	0.036	15.503	15.500	0.003	0.490	0 1	0
3	A151086	151086	15.620	0.039	15.658	16.000	–	0.490	0 0	10
3	151086	A151107	15.908	0.010	16.000	16.000	–	0.490	0 1	0
3	A151107	151107	16.117	0.041	16.158	16.500	–	0.490	0 0	10
3	151107	A151116	16.341	0.000	16.717	16.717	–	0.490	0 1	0
3	A151116	151116	16.833	0.055	16.888	17.217	–	0.490	0 0	11
3	151116	A151117	16.972	0.000	17.000	17.000	–	0.490	0 1	0
3	A151117	151117	17.150	0.016	17.166	17.500	–	0.490	0 0	10
3	151117	A151119	17.249	0.030	17.500	17.500	–	0.490	0 1	0
3	A151119	151119	17.600	0.039	17.639	18.000	–	0.490	0 0	9
3	151119	A151121	17.739	0.056	17.795	17.750	0.045	0.534	1 0	0
3	A151121	A151123	17.795	0.048	17.843	17.833	0.009	0.544	1 1	0
3	A151123	A151125	17.993	0.062	18.055	18.017	0.038	0.582	3 1	0
3	A151125	151125	18.105	0.045	18.150	18.517	–	0.582	1 1	6
3	151125	151121	18.150	0.055	18.205	18.250	–	0.582	0 1	25
3	151121	151123	18.205	0.009	18.214	18.333	–	0.582	0 0	23
3	151123	A151127	18.247	0.065	18.400	18.400	–	0.582	1 0	0
3	A151127	A151128	18.433	0.053	18.500	18.500	–	0.582	1 1	0
3	A151128	151127	18.550	0.073	18.623	18.900	–	0.582	0 1	14
3	151127	151128	18.623	0.039	18.661	19.000	–	0.582	0 0	10
3	151128	A151131	18.811	0.045	19.000	19.000	–	0.582	0 1	0
3	A151131	151131	19.117	0.053	19.170	19.500	–	0.582	0 0	11
3	151131	A151133	19.270	0.034	19.533	19.533	–	0.582	1 0	0
3	A151133	A151135	19.533	0.056	19.867	19.867	–	0.582	1 0	0
3	A151135	151133	19.883	0.000	19.883	20.033	–	0.582	0 0	22
3	151133	151135	19.883	0.035	19.918	20.367	–	0.582	0 0	4
3	151135	A151136	19.918	0.058	20.250	20.250	–	0.582	0 0	0
3	A151136	A151137	20.250	0.000	20.383	20.383	–	0.582	1 0	0
3	A151137	151136	20.383	0.041	20.424	20.750	–	0.582	1 0	11
3	151136	151137	20.424	0.076	20.500	20.883	–	0.582	0 0	8
3	151137	A151138	20.500	0.057	20.733	20.733	–	0.582	0 1	0
3	A151138	151138	20.900	0.053	20.953	21.233	–	0.582	0 0	14
3	151138	A151144	21.036	0.054	21.933	21.933	–	0.582	2 0	0
3	A151144	151144	21.983	0.019	22.002	22.433	–	0.582	0 0	5
3	151144	A151145	22.102	0.025	22.317	22.317	–	0.582	0 0	0
3	A151145	A151146	22.367	0.136	22.503	22.417	0.086	0.668	3 1	0
3	A151146	151146	22.503	0.003	22.506	22.917	–	0.668	0 0	1
3	151146	A151147	22.613	0.026	22.683	22.683	–	0.668	0 0	0
3	A151147	151147	22.783	0.023	22.807	23.183	–	0.668	0 0	8
3	151147	151145	22.807	0.000	22.807	22.817	–	0.668	0 0	30
3	151145	A151148	22.807	0.087	23.000	23.000	–	0.668	0 0	0
3	A151148	A151149	23.017	0.037	23.053	23.000	0.053	0.721	0 0	0
3	A151149	151149	23.087	0.014	23.100	23.500	–	0.721	0 0	3
3	151149	A151150	23.134	0.076	23.250	23.250	–	0.721	1 0	0
3	A151150	151148	23.267	0.025	23.292	23.500	–	0.721	1 0	18

3	151148	151150	23.292	0.000	23.292	23.750	–	0.721	0 0	3
3	151150	DD1	23.292	0.026	23.318	24.000	–	0.721	0 0	3
4	DS1	A150847	7.500	0.037	7.537	7.500	0.037	0.037	0 1	0
4	A150847	150847	7.570	0.039	7.609	8.000	–	0.037	0 0	5
4	150847	A150842	7.759	0.014	7.773	7.750	0.023	0.060	0 1	0
4	A150842	150842	7.840	0.081	7.921	8.250	–	0.060	0 0	9
4	150842	A150887	8.004	0.067	8.071	8.000	0.071	0.131	1 0	0
4	A150887	A150890	8.121	0.052	8.173	8.000	0.173	0.304	2 0	0
4	A150890	A150877	8.406	0.071	8.477	8.250	0.227	0.531	3 0	0
4	A150877	150890	8.477	0.045	8.522	8.500	0.022	0.553	2 0	21
4	150890	A150913	8.539	0.020	8.558	8.350	0.208	0.761	2 1	0
4	A150913	150877	8.658	0.028	8.686	8.750	–	0.761	1 1	13
4	150877	150887	8.686	0.000	8.686	8.500	0.186	0.947	0 1	37
4	150887	150913	8.770	0.065	8.835	8.850	–	0.947	0 0	17
4	150913	A150876	9.452	0.043	9.500	9.500	–	0.947	0 1	0
4	A150876	150876	9.600	0.057	9.657	10.000	–	0.947	0 0	10
4	150876	A150948	9.691	0.002	9.693	9.667	0.026	0.974	0 1	0
4	A150948	A150952	9.910	0.047	9.956	9.733	0.223	1.197	5 1	0
4	A150952	A150957	9.956	0.052	10.008	10.000	0.008	1.205	5 1	0
4	A150957	150948	10.075	0.019	10.093	10.167	–	1.205	5 0	25
4	150948	150957	10.210	0.011	10.221	10.500	–	1.205	5 0	13
4	150957	150952	10.221	0.044	10.264	10.233	0.031	1.236	0 0	19
4	150952	A150973	10.264	0.060	10.324	10.183	0.141	1.377	1 0	0
4	A150973	A150966	10.408	0.000	10.408	10.250	0.158	1.535	3 0	0
4	A150966	150973	10.524	0.081	10.605	10.683	–	1.535	2 0	17
4	150973	A150987	10.605	0.056	10.661	10.600	0.061	1.596	2 1	0
4	A150987	150966	10.695	0.073	10.767	10.750	0.017	1.613	0 1	22
4	150966	150987	10.834	0.038	10.872	11.100	–	1.613	0 0	13
4	150987	A150995	10.955	0.036	10.991	10.983	0.008	1.621	0 1	0
4	A150995	150995	11.141	0.065	11.207	11.483	–	1.621	0 0	13
4	150995	A151011	11.290	0.063	11.483	11.483	–	1.621	0 1	0
4	A151011	151011	11.533	0.078	11.611	11.983	–	1.621	0 0	8
4	151011	A151014	11.678	0.052	11.730	11.650	0.080	1.702	0 1	0
4	A151014	151014	11.880	0.074	11.955	12.150	–	1.702	0 0	14
4	151014	A16	12.071	0.041	12.112	12.000	0.112	1.814	5 1	0
4	A16	16	12.245	0.000	12.245	12.500	–	1.814	0 0	8
4	16	A151032	12.329	0.046	12.383	12.383	–	1.814	0 1	0
4	A151032	151032	12.583	0.057	12.640	12.883	–	1.814	0 0	16
4	151032	A151024	12.740	0.000	12.740	12.500	0.240	2.053	1 1	0
4	A151024	151024	12.923	0.057	12.980	13.000	–	2.053	0 0	15
4	151024	A151051	13.047	0.029	13.076	13.067	0.009	2.063	0 1	0
4	A151051	151051	13.326	0.016	13.342	13.567	–	2.063	0 0	16
4	151051	A151022	13.425	0.030	13.455	13.250	0.205	2.268	0 1	0
4	A151022	151022	13.605	0.042	13.647	13.750	–	2.268	0 0	12
4	151022	DD1	13.747	0.074	13.821	24.000	–	2.268	0 0	12
5	DS1	A150870	7.987	0.013	8.000	8.000	–	0.000	0 1	0
5	A150870	150870	8.067	0.029	8.095	8.500	–	0.000	0 0	6
5	150870	A150908	8.179	0.066	8.250	8.250	–	0.000	0 1	0
5	A150908	150908	8.467	0.014	8.480	8.750	–	0.000	0 0	14
5	150908	A150871	8.547	0.073	8.620	8.500	0.120	0.120	0 1	0
5	A150871	A150923	8.704	0.029	8.732	8.717	0.016	0.136	1 1	0
5	A150923	150871	8.749	0.073	8.822	9.000	–	0.136	1 0	13
5	150871	A150930	8.889	0.067	8.956	8.950	0.006	0.142	2 0	0
5	A150930	A150924	8.956	0.033	8.989	8.833	0.155	0.297	2 1	0
5	A150924	150923	9.139	0.014	9.152	9.217	–	0.297	1 1	26

5	150923	150930	9.169	0.057	9.226	9.450	–	0.297	0	1	17
5	150930	150924	9.226	0.019	9.245	9.333	–	0.297	0	0	16
5	150924	A150941	9.345	0.020	9.417	9.417	–	0.297	1	0	0
5	A150941	A150943	9.433	0.060	9.500	9.500	–	0.297	1	1	0
5	A150943	150943	9.633	0.046	9.680	10.000	–	0.297	1	0	11
5	150943	150941	9.763	0.081	9.844	9.917	–	0.297	0	0	26
5	150941	A150958	9.844	0.000	9.844	9.817	0.027	0.325	0	1	0
5	A150958	150958	9.927	0.074	10.002	10.317	–	0.325	0	0	10
5	150958	A150971	10.152	0.063	10.215	10.150	0.065	0.389	0	1	0
5	A150971	150971	10.298	0.078	10.376	10.650	–	0.389	0	0	10
5	150971	A150984	10.409	0.039	10.448	10.433	0.015	0.404	1	0	0
5	A150984	150984	10.448	0.075	10.523	10.933	–	0.404	0	0	5
5	150984	A150985	10.523	0.016	10.539	10.517	0.022	0.426	2	0	0
5	A150985	A150986	10.755	0.085	10.840	10.617	0.224	0.650	4	1	0
5	A150986	150985	10.874	0.044	10.917	11.017	–	0.650	2	1	23
5	150985	150986	10.984	0.087	11.071	11.117	–	0.650	0	0	14
5	150986	A150994	11.188	0.032	11.219	11.200	0.019	0.669	0	1	0
5	A150994	A150840	11.419	0.076	11.500	11.500	–	0.669	1	1	0
5	A150840	150994	11.500	0.000	11.500	11.250	0.250	0.919	1	0	17
5	150994	150840	11.600	0.082	11.682	12.000	–	0.919	0	0	11
5	150840	A151019	11.682	0.040	11.750	11.750	–	0.919	0	1	0
5	A151019	151019	11.800	0.074	11.874	12.250	–	0.919	0	0	8
5	151019	A11	11.941	0.041	12.000	12.000	–	0.919	5	1	0
5	A11	11	12.383	0.000	12.383	12.500	–	0.919	0	0	23
5	11	A5	12.633	0.000	12.633	12.500	0.133	1.052	5	1	0
5	A5	5	12.767	0.000	12.767	13.000	–	1.052	0	0	8
5	5	A151041	12.917	0.061	12.978	12.750	0.228	1.280	2	0	0
5	A151041	A151045	13.011	0.055	13.083	13.083	–	1.280	2	1	0
5	A151045	151041	13.217	0.021	13.237	13.250	–	1.280	0	1	16
5	151041	151045	13.287	0.045	13.333	13.583	–	1.280	0	0	15
5	151045	A151083	13.433	0.036	14.617	14.617	–	1.280	1	0	0
5	A151083	151083	14.850	0.087	14.937	15.117	–	1.280	0	0	20
5	151083	DD1	14.937	0.047	14.984	24.000	–	1.280	0	0	20
6	DS1	A150898	7.904	0.013	7.917	7.917	–	0.000	0	1	0
6	A150898	150898	7.950	0.075	8.025	8.417	–	0.000	0	0	7
6	150898	A150902	8.091	0.074	8.250	8.250	–	0.000	0	1	0
6	A150902	150902	8.367	0.015	8.382	8.750	–	0.000	0	0	8
6	150902	A150854	8.532	0.028	8.560	8.500	0.060	0.060	0	1	0
6	A150854	150854	8.677	0.055	8.732	9.000	–	0.060	0	0	11
6	150854	A150880	8.882	0.023	8.905	8.750	0.155	0.215	1	0	0
6	A150880	A150928	8.988	0.000	8.988	8.883	0.105	0.319	1	1	0
6	A150928	150880	9.138	0.075	9.213	9.250	–	0.319	0	1	19
6	150880	150928	9.229	0.087	9.317	9.383	–	0.319	0	0	20
6	150928	A150935	9.367	0.128	9.500	9.500	–	0.319	5	1	0
6	A150935	150935	9.617	0.003	9.620	10.000	–	0.319	0	0	8
6	150935	A150883	9.726	0.017	9.743	9.500	0.243	0.563	1	0	0
6	A150883	A150950	9.793	0.075	9.869	9.750	0.119	0.681	1	1	0
6	A150950	150950	10.019	0.060	10.079	10.250	–	0.681	1	0	13
6	150950	150883	10.162	0.004	10.166	10.000	0.166	0.847	0	0	26
6	150883	A150927	10.783	0.081	11.000	11.000	–	0.847	0	1	0
6	A150927	150927	11.200	0.060	11.260	11.500	–	0.847	0	0	16
6	150927	A151008	11.310	0.075	11.450	11.450	–	0.847	1	0	0
6	A151008	A150972	11.450	0.076	11.526	11.500	0.026	0.874	1	1	0
6	A150972	151008	11.643	0.039	11.682	11.950	–	0.874	0	1	14
6	151008	150972	11.682	0.000	11.682	12.000	–	0.874	0	0	10
6	150972	A151018	11.799	0.051	11.850	11.750	0.100	0.974	0	1	0

```
6 A151018 151018  11.883  0.074 11.958 12.250   -     0.974 0 0  7
6 151018  A151034 12.024  0.070 12.400 12.400   -     0.974 2 0  0
6 A151034 A151042 12.600  0.046 12.646 12.600  0.046  1.020 3 0  0
6 A151042 151034  12.679  0.000 12.679 12.900   -     1.020 1 0 17
6 151034  151042  12.796  0.077 12.873 13.100   -     1.020 0 0 14
6 151042  A151048 12.973  0.058 13.031 13.000  0.030  1.050 1 0  0
6 A151048 A151047 13.081  0.031 13.167 13.167   -     1.050 1 1  0
6 A151047 151048  13.300  0.020 13.320 13.500   -     1.050 0 1 18
6 151048  151047  13.320  0.029 13.349 13.667   -     1.050 0 0 11
6 151047  A151081 13.399  0.076 14.550 14.550   -     1.050 1 1  0
6 A151081 151081  14.633  0.057 14.690 15.050   -     1.050 0 0  9
6 151081  A151095 14.807  0.036 15.500 15.500   -     1.050 0 0  0
6 A151095 151095  15.583  0.041 15.624 16.000   -     1.050 0 0  8
6 151095  DD1     15.691  0.078 15.769 24.000   -     1.050 0 0  8

7 DS1      A150846  8.000  0.037  8.037  8.000  0.037  0.037 0 1  0
7 A150846  A150901  8.237  0.000  8.250  8.250   -     0.037 1 1  0
7 A150901  A150900  8.250  0.000  8.250  8.250   -     0.037 2 1  0
7 A150900  150846   8.267  0.040  8.307  8.500   -     0.037 2 0 17
7 150846   A150916  8.373  0.067  8.440  8.433  0.007  0.044 3 0  0
7 A150916  150916   8.457  0.040  8.497  8.933   -     0.044 2 0  4
7 150916   A150799  8.497  0.024  8.521  8.500  0.021  0.065 2 1  0
7 A150799  150900   8.654  0.070  8.724  8.750   -     0.065 1 1 29
7 150900   150901   8.758  0.000  8.758  8.750  0.008  0.072 0 1 31
7 150901   A150844  8.858  0.002  8.859  8.750  0.109  0.182 2 1  0
7 A150844  150799   9.009  0.081  9.091  9.000  0.091  0.272 2 0 35
7 150799   A150891  9.124  0.000  9.124  9.000  0.124  0.396 2 1  0
7 A150891  150844   9.174  0.097  9.271  9.250  0.021  0.417 0 1 25
7 150844   150891   9.354  0.000  9.354  9.500   -     0.417 0 0 14
7 150891   A150944  9.454  0.055  9.509  9.483  0.026  0.443 0 1  0
7 A150944  A150945  9.576  0.023  9.599  9.583  0.016  0.459 1 1  0
7 A150945  150944   9.683  0.075  9.757  9.983   -     0.459 1 0 15
7 150944   A150955  9.807  0.000  9.807  9.800  0.007  0.466 1 1  0
7 A150955  150955   9.974  0.067 10.041 10.300   -     0.466 1 0 15
7 150955   150945  10.141  0.074 10.215 10.083  0.132  0.598 0 0 37
7 150945   A150974 10.232  0.000 10.232 10.200  0.032  0.629 0 1  0
7 A150974  A150969 10.298  0.026 10.324 10.250  0.074  0.704 2 1  0
7 A150969  A150979 10.441  0.010 10.451 10.383  0.068  0.771 3 1  0
7 A150979  150974  10.534  0.076 10.611 10.700   -     0.771 3 0 23
7 150974   150969  10.711  0.057 10.768 10.750  0.018  0.789 1 0 27
7 150969   150979  10.868  0.029 10.897 10.883  0.014  0.803 0 0 27
7 150979   A150996 10.897  0.021 11.000 11.000   -     0.803 2 0  0
7 A150996  A150850 11.233  0.053 11.286 11.250  0.036  0.839 3 1  0
7 A150850  150850  11.336  0.057 11.393 11.750   -     0.839 2 0  7
7 150850   A151010 11.460  0.055 11.515 11.483  0.032  0.871 3 0  0
7 A151010  150996  11.532  0.000 11.532 11.500  0.032  0.903 1 0 32
7 150996   A151009 11.582  0.039 11.621 11.500  0.121  1.024 1 1  0
7 A151009  151010  11.655  0.076 11.731 11.983   -     1.024 0 1 13
7 151010   151009  11.731  0.048 11.778 12.000   -     1.024 0 0 10
7 151009   A151023 11.845  0.022 11.867 11.817  0.050  1.075 0 1  0
7 A151023  A150879 11.917  0.075 11.992 11.750  0.242  1.317 1 1  0
7 A150879  150879  12.159  0.052 12.210 12.250   -     1.317 0 1 14
7 150879   151023  12.210  0.076 12.287 12.317   -     1.317 0 0 26
7 151023   A12     12.353  0.042 12.500 12.500   -     1.317 5 1  0
7 A12      12      12.533  0.000 12.533 13.000   -     1.317 0 0  2
7 12       A151033 12.600  0.042 12.642 12.500  0.142  1.458 0 1  0
7 A151033  151033  12.842  0.067 12.909 13.000   -     1.458 0 0 17
```

7	151033	A151040	12.926	0.014	12.939	12.583	0.356	1.814	0 1	0
7	A151040	151040	13.189	0.039	13.228	13.083	0.144	1.959	0 0	18
7	151040	A151013	13.261	0.000	13.261	13.250	0.011	1.970	0 1	0
7	A151013	151013	13.394	0.072	13.466	13.750	–	1.970	0 0	13
7	151013	A151065	13.600	0.060	13.850	13.850	–	1.970	0 1	0
7	A151065	151065	13.883	0.043	13.926	14.350	–	1.970	0 0	5
7	151065	A151080	14.176	0.070	14.483	14.483	–	1.970	0 1	0
7	A151080	151080	14.633	0.078	14.711	14.983	–	1.970	0 0	14
7	151080	A151094	14.778	0.030	15.500	15.500	–	1.970	2 0	0
7	A151094	151094	15.633	0.039	15.672	16.000	–	1.970	0 0	11
7	151094	DD1	15.739	0.053	15.792	24.000	–	1.970	0 0	11
8	DS1	A150857	8.000	0.037	8.037	8.000	0.037	0.037	0 1	0
8	A150857	A150838	8.153	0.046	8.250	8.250	–	0.037	1 1	0
8	A150838	150857	8.317	0.021	8.338	8.500	–	0.037	1 0	19
8	150857	A150918	8.471	0.000	8.471	8.450	0.021	0.058	2 0	0
8	A150918	150838	8.555	0.028	8.583	8.750	–	0.058	1 0	20
8	150838	A150853	8.583	0.000	8.583	8.500	0.083	0.141	1 1	0
8	A150853	150853	8.833	0.032	8.865	9.000	–	0.141	1 0	17
8	150853	A150926	8.949	0.004	8.952	8.800	0.152	0.293	2 0	0
8	A150926	A150882	8.952	0.000	9.000	9.000	–	0.293	3 0	0
8	A150882	150918	9.000	0.076	9.076	8.950	0.126	0.420	2 0	37
8	150918	150926	9.076	0.087	9.163	9.300	–	0.420	1 0	13
8	150926	A150885	9.263	0.037	9.300	9.250	0.050	0.469	1 1	0
8	A150885	150882	9.383	0.039	9.422	9.500	–	0.469	0 1	26
8	150882	150885	9.422	0.043	9.465	9.750	–	0.469	0 0	10
8	150885	A150937	9.548	0.065	9.614	9.500	0.114	0.583	0 1	0
8	A150937	150937	9.814	0.040	9.854	10.000	–	0.583	0 0	15
8	150937	A150949	9.904	0.076	9.979	9.750	0.229	0.812	0 1	0
8	A150949	A150968	10.196	0.040	10.236	10.050	0.186	0.998	1 1	0
8	A150968	150949	10.286	0.039	10.325	10.250	0.075	1.073	1 0	21
8	150949	150968	10.425	0.065	10.490	10.550	–	1.073	0 0	16
8	150968	A150980	10.557	0.046	10.603	10.400	0.203	1.276	0 1	0
8	A150980	150980	10.686	0.045	10.731	10.900	–	1.276	0 0	8
8	150980	A150988	10.798	0.065	10.863	10.750	0.113	1.389	2 0	0
8	A150988	A150992	10.996	0.000	11.000	11.000	–	1.389	3 1	0
8	A150992	150988	11.117	0.046	11.163	11.250	–	1.389	1 1	18
8	150988	150992	11.246	0.021	11.267	11.500	–	1.389	0 0	17
8	150992	A151007	11.317	0.085	11.402	11.350	0.052	1.441	1 1	0
8	A151007	151007	11.536	0.044	11.579	11.850	–	1.441	0 0	11
8	151007	A151016	11.646	0.058	11.704	11.700	0.004	1.445	0 1	0
8	A151016	151016	11.788	0.075	11.863	12.200	–	1.445	0 0	10
8	151016	A151027	11.946	0.058	12.250	12.250	–	1.445	0 1	0
8	A151027	151027	12.417	0.074	12.491	12.750	–	1.445	0 0	15
8	151027	A151039	12.624	0.076	12.750	12.750	–	1.445	0 1	0
8	A151039	151039	12.800	0.047	12.847	13.250	–	1.445	0 0	6
8	151039	A150982	12.897	0.000	12.897	12.667	0.230	1.675	0 1	0
8	A150982	150982	12.980	0.018	12.999	13.167	–	1.675	0 0	7
8	150982	A151053	13.165	0.057	13.222	13.100	0.122	1.797	2 0	0
8	A151053	A151052	13.339	0.013	13.351	13.250	0.101	1.898	2 1	0
8	A151052	151053	13.384	0.081	13.465	13.600	–	1.898	0 1	15
8	151053	151052	13.548	0.033	13.581	13.750	–	1.898	0 0	14
8	151052	A151066	13.681	0.040	14.000	14.000	–	1.898	1 0	0
8	A151066	151066	14.050	0.081	14.131	14.500	–	1.898	0 0	8
8	151066	A151074	14.148	0.058	14.383	14.383	–	1.898	0 1	0
8	A151074	151074	14.433	0.078	14.511	14.883	–	1.898	0 0	8
8	151074	A151084	14.595	0.036	14.667	14.667	–	1.898	0 1	0

```
8 A151084  151084   14.817  0.026 14.843 15.167    -      1.898 0 0 11
8 151084   A151099  14.926  0.064 15.400 15.400    -      1.898 0 1  0
8 A151099  151099   15.567  0.041 15.608 15.900    -      1.898 0 0 13
8 151099   A151108  15.691  0.053 16.067 16.067    -      1.898 0 1  0
8 A151108  151108   16.150  0.023 16.173 16.567    -      1.898 0 0  7
8 151108   DD1      16.257  0.072 16.328 24.000    -      1.898 0 0  7

9 DS1      A150886   8.000  0.037  8.037  8.000  0.037   0.037 0 1  0
9 A150886  150886    8.070  0.044  8.114  8.500    -     0.037 0 0  5
9 150886   A150910   8.231  0.020  8.250  8.217  0.034   0.070 1 0  0
9 A150910  A150915   8.484  0.000  8.484  8.400  0.084   0.154 2 0  0
9 A150915  150910    8.584  0.065  8.649  8.717    -     0.154 1 0 24
9 150910   A150881   8.766  0.000  8.766  8.750  0.016   0.170 2 0  0
9 A150881  A150892   8.766  0.000  8.766  8.750  0.016   0.186 2 1  0
9 A150892  150915    8.849  0.000  8.849  8.900    -     0.186 1 1 22
9 150915   A150929   8.849  0.017  9.000  9.000    -     0.186 2 1  0
9 A150929  150929    9.000  0.056  9.056  9.500    -     0.186 1 1  4
9 150929   150881    9.056  0.041  9.098  9.250    -     0.186 0 1 20
9 150881   150892    9.098  0.000  9.098  9.250    -     0.186 0 0 20
9 150892   A150938   9.198  0.074  9.300  9.300    -     0.186 0 1  0
9 A150938  A150750   9.517  0.075  9.591  9.500  0.091   0.277 1 1  0
9 A150750  150938    9.591  0.067  9.659  9.800    -     0.277 1 0 22
9 150938   150750    9.742  0.036  9.778 10.000    -     0.277 0 0 12
9 150750   A150934   9.862  0.026  9.888  9.750  0.138   0.415 0 1  0
9 A150934  150934    9.971  0.009  9.980 10.250    -     0.415 0 0  6
9 150934   A150961  10.013  0.028 10.042 10.000  0.042   0.457 1 0  0
9 A150961  A150960  10.042  0.076 10.118  9.867  0.251   0.707 2 1  0
9 A150960  150960   10.318  0.057 10.374 10.367  0.007   0.715 1 0 16
9 150960   150961   10.491  0.017 10.508 10.500  0.008   0.723 0 0 29
9 150961   A150993  10.525  0.038 10.950 10.950    -     0.723 1 0  0
9 A150993  A150997  11.000  0.000 11.050 11.050    -     0.723 2 0  0
9 A150997  150997   11.050  0.042 11.092 11.550    -     0.723 1 0  3
9 150997   150993   11.092  0.030 11.122 11.450    -     0.723 0 0 11
9 150993   A151005  11.189  0.011 11.283 11.283    -     0.723 1 1  0
9 A151005  151005   11.333  0.057 11.390 11.783    -     0.723 0 0  7
9 151005   A151006  11.457  0.069 11.526 11.283  0.242   0.965 5 0  0
9 A151006  A150911  11.609  0.061 11.670 11.500  0.170   1.135 5 1  0
9 A150911  151006   11.786  0.043 11.829 11.783  0.046   1.181 0 1 19
9 151006   150911   11.846  0.062 11.908 12.000    -     1.181 0 0 15
9 150911   A151035  11.958  0.074 12.483 12.483    -     1.181 0 1  0
9 A151035  151035   12.533  0.042 12.575 12.983    -     1.181 0 0  6
9 151035   A151038  12.692  0.039 12.731 12.517  0.214   1.395 0 1  0
9 A151038  151038   13.031  0.084 13.114 13.017  0.098   1.493 0 0 24
9 151038   A151055  13.214  0.068 13.282 13.250  0.032   1.525 0 1  0
9 A151055  151055   13.315  0.058 13.373 13.750    -     1.525 0 0  6
9 151055   A151064  13.490  0.053 13.800 13.800    -     1.525 0 1  0
9 A151064  151064   14.000  0.045 14.045 14.300    -     1.525 0 0 15
9 151064   A151072  14.129  0.052 14.283 14.283    -     1.525 1 0  0
9 A151072  A151078  14.350  0.016 14.450 14.450    -     1.525 1 1  0
9 A151078  151072   14.667  0.073 14.740 14.783    -     1.525 0 1 28
9 151072   151078   14.740  0.077 14.816 14.950    -     1.525 0 0 22
9 151078   A151089  14.883  0.016 15.017 15.017    -     1.525 1 0  0
9 A151089  151089   15.083  0.075 15.159 15.517    -     1.525 0 0  9
9 151089   A151096  15.159  0.056 15.367 15.367    -     1.525 0 1  0
9 A151096  151096   15.467  0.078 15.545 15.867    -     1.525 0 0 11
9 151096   A151103  15.612  0.033 16.000 16.000    -     1.525 0 1  0
9 A151103  151103   16.250  0.020 16.270 16.500    -     1.525 0 0 17
```

9	151103	DD1	16.436	0.026	16.463	24.000	–	1.525	0	0	17
10	DS1	A150921	8.550	0.017	8.567	8.567	–	0.000	1	0	0
10	A150921	150921	8.567	0.058	8.625	9.067	–	0.000	0	0	4
10	150921	A150919	8.625	0.041	8.750	8.750	–	0.000	1	0	0
10	A150919	A150893	8.750	0.074	8.824	8.750	0.074	0.074	1	1	0
10	A150893	150919	8.991	0.015	9.006	9.250	–	0.074	0	1	16
10	150919	A150751	9.006	0.073	9.079	9.000	0.079	0.153	2	1	0
10	A150751	150893	9.162	0.000	9.162	9.250	–	0.153	2	0	21
10	150893	A150897	9.312	0.082	9.500	9.500	–	0.153	2	1	0
10	A150897	150751	9.533	0.009	9.542	9.500	0.042	0.196	1	1	28
10	150751	A150953	9.692	0.052	9.767	9.767	–	0.196	1	1	0
10	A150953	150897	9.767	0.014	9.780	10.000	–	0.196	1	0	17
10	150897	A150954	9.847	0.073	10.000	10.000	–	0.196	1	1	0
10	A150954	150953	10.033	0.027	10.060	10.267	–	0.196	0	1	18
10	150953	150954	10.060	0.063	10.123	10.500	–	0.196	0	0	8
10	150954	A150975	10.190	0.004	10.200	10.200	–	0.196	0	1	0
10	A150975	150975	10.250	0.026	10.276	10.700	–	0.196	0	0	5
10	150975	A150977	10.360	0.009	10.368	10.250	0.118	0.314	0	1	0
10	A150977	150977	10.635	0.009	10.644	10.750	–	0.314	0	0	17
10	150977	A150991	10.777	0.082	10.883	10.883	–	0.314	0	1	0
10	A150991	A150998	10.967	0.000	11.067	11.067	–	0.314	1	1	0
10	A150998	A150999	11.150	0.053	11.203	11.117	0.086	0.400	3	1	0
10	A150999	150998	11.586	0.023	11.609	11.567	0.042	0.442	2	1	33
10	150998	150991	11.609	0.024	11.633	11.383	0.250	0.692	2	0	45
10	150991	A151012	11.700	0.000	11.700	11.567	0.133	0.825	2	1	0
10	A151012	150999	11.816	0.059	11.875	11.617	0.259	1.083	0	1	41
10	150999	151012	12.125	0.051	12.176	12.067	0.109	1.193	0	0	29
10	151012	A151031	12.292	0.064	12.367	12.367	–	1.193	0	1	0
10	A151031	151031	12.400	0.084	12.484	12.867	–	1.193	0	0	8
10	151031	A8	12.584	0.037	12.621	12.500	0.121	1.314	5	1	0
10	A8	8	12.771	0.000	12.771	13.000	–	1.314	0	0	10
10	8	A151037	12.821	0.040	12.861	12.750	0.111	1.425	1	0	0
10	A151037	151037	12.911	0.029	12.940	13.250	–	1.425	0	0	5
10	151037	A1	12.940	0.070	13.010	13.000	0.010	1.435	5	1	0
10	A1	1	13.144	0.000	13.144	13.500	–	1.435	0	0	8
10	1	A151056	13.194	0.038	13.231	13.200	0.031	1.466	0	1	0
10	A151056	151056	13.281	0.074	13.355	13.700	–	1.466	0	0	8
10	151056	A151067	13.605	0.055	14.000	14.000	–	1.466	0	1	0
10	A151067	151067	14.067	0.066	14.133	14.500	–	1.466	0	0	8
10	151067	A151069	14.216	0.025	14.250	14.250	–	1.466	1	0	0
10	A151069	151069	14.300	0.058	14.358	14.750	–	1.466	0	0	7
10	151069	A151077	14.358	0.017	14.433	14.433	–	1.466	0	1	0
10	A151077	A151082	14.717	0.017	14.750	14.750	–	1.466	1	1	0
10	A151082	151077	14.750	0.056	14.806	14.933	–	1.466	1	0	23
10	151077	151082	14.922	0.029	14.952	15.250	–	1.466	0	0	13
10	151082	A151091	14.985	0.042	15.167	15.167	–	1.466	0	1	0
10	A151091	151091	15.283	0.053	15.336	15.667	–	1.466	0	0	11
10	151091	A151101	15.420	0.014	15.500	15.500	–	1.466	0	1	0
10	A151101	151101	15.600	0.062	15.662	16.000	–	1.466	0	0	10
10	151101	A151104	15.762	0.007	15.833	15.833	–	1.466	0	1	0
10	A151104	151104	15.950	0.077	16.027	16.333	–	1.466	0	0	12
10	151104	A151112	16.127	0.039	16.583	16.583	–	1.466	0	1	0
10	A151112	151112	16.717	0.053	16.770	17.083	–	1.466	0	0	12
10	151112	DD1	16.903	0.069	16.971	24.000	–	1.466	0	0	12
11	DS1	A150917	8.691	0.059	8.750	8.750	–	0.000	1	0	0

```
11 A150917 150917   8.800 0.058  8.858  9.250   -   0.000 0 0  7
11 150917  A150947  9.475 0.000  9.600  9.600   -   0.000 0 1  0
11 A150947 A150951  9.650 0.043  9.750  9.750   -   0.000 2 1  0
11 A150951 150947   9.800 0.042  9.842 10.100   -   0.000 2 0 15
11 150947  A150963  9.875 0.030 10.000 10.000   -   0.000 2 1  0
11 A150963 150951  10.083 0.040 10.123 10.250   -   0.000 0 1 23
11 150951  150963  10.223 0.073 10.297 10.500   -   0.000 0 0 18
11 150963  A150990 10.363 0.028 10.750 10.750   -   0.000 0 1  0
11 A150990 150990  10.817 0.076 10.893 11.250   -   0.000 0 0  9
11 150990  A151001 10.976 0.000 11.167 11.167   -   0.000 0 1  0
11 A151001 151001  11.233 0.077 11.310 11.667   -   0.000 0 0  9
11 151001  A10     11.427 0.042 11.500 11.500   -   0.000 5 1  0
11 A10     10      11.600 0.000 11.600 12.000   -   0.000 0 0  6
11 10      A151021 11.733 0.024 12.000 12.000   -   0.000 0 1  0
11 A151021 151021  12.050 0.081 12.131 12.500   -   0.000 0 0  8
11 151021  A151043 12.198 0.086 12.383 12.383   -   0.000 1 0  0
11 A151043 A151036 12.450 0.024 12.483 12.483   -   0.000 1 1  0
11 A151036 151043  12.700 0.094 12.794 12.883   -   0.000 0 1 25
11 151043  151036  12.794 0.029 12.823 12.983   -   0.000 0 0 21
11 151036  A151050 12.956 0.016 13.067 13.067   -   0.000 0 1  0
11 A151050 151050  13.100 0.039 13.139 13.567   -   0.000 0 0  5
11 151050  A151060 13.222 0.081 13.500 13.500   -   0.000 3 1  0
11 A151060 151060  13.650 0.074 13.724 14.000   -   0.000 0 0 14
11 151060  A151063 13.874 0.069 14.000 14.000   -   0.000 1 0  0
11 A151063 A151071 14.017 0.051 14.133 14.133   -   0.000 2 0  0
11 A151071 151071  14.183 0.038 14.222 14.633   -   0.000 1 0  6
11 151071  151063  14.222 0.061 14.282 14.500   -   0.000 0 0 17
11 151063  A151076 14.282 0.024 14.433 14.433   -   0.000 0 1  0
11 A151076 151076  14.567 0.074 14.641 14.933   -   0.000 0 0 13
11 151076  A150848 14.741 0.003 14.750 14.750   -   0.000 0 1  0
11 A150848 150848  14.950 0.054 15.004 15.250   -   0.000 0 0 16
11 150848  A151093 15.171 0.050 15.250 15.250   -   0.000 0 1  0
11 A151093 151093  15.283 0.024 15.307 15.750   -   0.000 0 0  4
11 151093  A151102 15.357 0.070 15.667 15.667   -   0.000 0 1  0
11 A151102 A151106 15.817 0.081 15.983 15.983   -   0.000 1 1  0
11 A151106 151102  15.983 0.033 16.016 16.167   -   0.000 1 0 21
11 151102  151106  16.150 0.033 16.183 16.483   -   0.000 0 0 12
11 151106  A151110 16.183 0.013 16.233 16.233   -   0.000 0 1  0
11 A151110 151110  16.367 0.022 16.389 16.733   -   0.000 0 0 10
11 151110  A151115 16.522 0.066 16.750 16.750   -   0.000 0 1  0
11 A151115 151115  16.867 0.075 16.941 17.250   -   0.000 0 0 12
11 151115  DD1     17.008 0.078 17.086 24.000   -   0.000 0 0 12
```

References

[1] E. H. L. Aarts and J. H. M. Korst. *Simulated Annealing and Boltzmann Machines.* Wiley, Chichester, UK, 1989.

[2] E. H. L. Aarts, J. H. M. Korst, and P. J. M. van Laarhoven. Simulated Annealing. In E. H. L. Aarts and J. K. Lenstra, editors, *Local Search in Combinatorial Optimization*, pages 91–120. Wiley, Chichester, UK, 1997.

[3] S. Albers. *The Influence of Lookahead in Competitive Online Algorithms.* Phd thesis, University of Saarland, Saarbrücken, 1993.

[4] S. Albers. Online Algorithms: A Survey. *Mathematical Programming series B*, 97:3–26, 2003.

[5] N. Ascheuer, M. Fischetti, and M. Grötschel. Solving the Asymmetric Travelling Salesman Problem with Time Windows by Branch-and-Cut. *Mathematical Programming*, Ser. A 90:475–506, 2001.

[6] P. Augerat, J. Belenguer, E. Benavent, A. Corberán, D. Naddef, and G. Rinaldi. Computational Results with a Branch and Cut Code for the Capacitated Vehicle Routing Problem. Technical Report RR 949-M, Université Joseph Fourier, Grenoble, France, 1995.

[7] B. M. Baker and M. A. Ayechew. A Genetic Algorithm for the Vehicle Routing Problem. *Computer and Operations Research*, 30, 2003.

[8] R. Baldacci, A. Mingozzi, and E. Hadjiconstantinou. An Exact Algorithm for the Capacitated Vehicle Routing Problem based on a Two-Commodity Network Flow Formulation. Technical Report 16, Departament of Mathematics, University of Bologna, Italy, 1999.

[9] C. Barnhart, E. L. Johnson, G. L. Nemhauser, M. W. P. Savelsberg, and P. H. Vance. Branch-and-Price: Column Generation for Solving Huge Integer Programs. *Operations Research*, 46(3):316–329, 1998.

[10] L. Bianco, A. Mingozzi, and S. Ricciardelli. An Exact Algorithm for Combining Vehicle Trips. In J. R. Daduna, L. Branco, and J. Paixao, editors, *Computer-Aided Transit Scheduling*, volume 430 of *Lecture Notes in Economics and Mathematical Systems*, pages 145–172. Springer, Berlin, 1995.

[11] A. Borodin and R. El-Yaniv. *Online Computation and Competitive Analysis.* Cambridge University Press, New York, USA, 1998.

[12] J. Bramel and D. Simchi-Levi. *The Logic of Logistics.* Springer, New York, USA, 1997.

[13] J. Brandao. Metaheuristic for the Vehicle Routing Problem with Time Windows. In S. Voss, S. Martello, I. H. Osman, and C. Roucairol, editors, *Meta Heuristics: Advances and Trends in Local Search Paradigms for Optimisation,* pages 19–36. Kluwer, Boston, MA, 1998.

[14] C. C. Carøe and R. Schultz. Dual Decomposition in Stochastic Integer Programming. *Operations Research Letters,* 24:37–45, 1999.

[15] V. Cerny. Thermodynamical Approach to the Traveling Salesman Problem: An Efficient Simulation Algorithm. *Opt. Theory Appl.,* 45(1):41–51, 1985.

[16] W.-C. Chiang and R. Russell. Simulated Annealing Metaheuristics for the Vehicle Routing Problem with Time Windows. *Annals of Operations Research,* 63:3–27, 1996.

[17] W.-C. Chiang and R. Russell. A Reactive Tabu Search Metaheuristic for the Vehicle Routing Problem with Time Windows. *INFORMS Journal on Computing,* 9:417–430, 1997.

[18] T. A. Ciriani, S. Gliozzi, E. L. Johnson, and R. Tadei. *Operational Research in Industry.* Macmillan, Houndmills, Basingstoke, UK, 1999.

[19] W. Cook and J. L. Rich. A Parallel Cutting Plane algorithm for the Vehicle Routing Problem with Time Windows. Technical report, Computational and Applied Mathematics, Rice University, Houston, TX, USA, 1999.

[20] J.-F. Cordeau, M. Gendreau, and G. Laporte. A Tabu Search Heuristic for Periodic and Multidepot Vehicle Routing Problem. *Networks,* 30:105–119, 1997.

[21] R. Cordone and R. W. Calvo. Note on Time Window Constraints in Routing Problems. Internal Report 96.005, Politecnico di Milano, Dipartimento di Elettronica e Informazione, Milan, Italy, 1996.

[22] N. Cwikla. *Patiententransporte in Krankenhäuser - Vergleich verschiedener Informationswege.* Diploma Thesis, Universität Kaiserslautern, Kaiserslautern, Germany, 2004.

[23] C. F. Daganzo. *Logistics System Analysis.* Springer, Berlin, Heidelberg, Germany, 2nd edition, 1996.

[24] B. Dantzig and P. Wolfe. The Decomposition Algorithm for Linear Programming. *Operations Research,* 8:101–111, 1960.

[25] G. Desaulniers, J. Desrosiers, A. Erdmann, M. M. Solomon, and F. Soumis. VRP with Pickup and Delivery. In P. Toth and D. Vigo, editors, *The Vehicle Routing Problem,* pages 225–242. Society for Industrial and Applied Mathematics, Philadelphia, 2001.

[26] M. Desrochers, J. Desrosiers, and M. M. Solomon. A New Optimization Algorithm for the Vehicle Routing Problem with Time Windows. *Operations Research*, 40(2):342–354, 1992, March-April.

[27] J. Desrosiers and M. E. Lübbecke. Selected Topics in Column Generation. *Operations Research*, 2004, submitted.

[28] R. W. Eglese. Simulated Annealing: A Tool for Operational Research. *European Journal of Operational Research*, 46:271–281, 1990.

[29] M. Esen. *Design, Implementation and Analysis of Online Bin Packing Algorithms*. Diploma thesis, University of Kaiserslautern, Kaiserslautern, September 2000.

[30] A. Fiat and G. J. Woeginger. *Online Algorithms - The State of the Art*. Springer, Berlin, Heidelberg, Germany, 1998.

[31] M. Fischetti, P. Toth, and D. Vigo. A Branch-and-Bound Algorithm for the Capacitated Vehicle Routing Problem on Directed Graphs. *Operations Research*, 42:846–859, 1994.

[32] M. R. Garey and D. S. Johnson. *Computers and Intractability - A Guide to the Theory of NP Completeness*. Freeman, New York, USA, 22nd edition, 2000.

[33] P. C. Gilmore and R. E. Gomory. A Linear Programming Approach to the Cutting Stock Problem, Part I. *Operations Research*, 9:849–859, 1961.

[34] P. C. Gilmore and R. E. Gomory. A Linear Programming Approach to the Cutting Stock Problem, Part II. *Operations Research*, 11:863–888, 1963.

[35] F. Glover and M. Laguna. *Tabu Search*. Kluwer Academic Publisher, Dordrecht, The Netherlands, 1997.

[36] B. L. Golden, E. A. Wasil, J. P. Kelly, and I. M. Chao. Metaheuristics in Vehicle Routing. In T. G. Crainic and G. Laporte, editors, *Fleet Management and Logistics*, pages 33–56. Kluwer, Boston, MA, USA, 1998.

[37] S. C. Graves, A. H. G. Rinnooy Kan, and P. H. Zipkin. *Logistics of Production and Inventory*. Elsevier, Amsterdam, The Netherlands, 1993.

[38] E. F. Grove. Online Bin Packing with Lookahead. In *Proceedings of the Sixth Annual ACM-SIAM Symposium on Discrete Algorithms*, pages 430–436, San Francisco, CA, 1995. ACM/SIAM.

[39] H. W. Hamacher, M. C. Müller, and S. Nickel. Modelling Rotastore - a Highly Parallel, Short Term Storage System. In *Operations Research Proceedings*, pages 513–521, Berlin, 1998. Springer Verlag.

[40] P. Hansen and N. Mladenovic. Variable Neighborhood Search: Principles and Applications. *EJOR*, 130:449–467, 2001.

[41] S. Heipcke. *Applications of Optimization with Xpress-MP*. Dash Optimization, Blisworth, UK, 2002.

[42] J. H. Holland. *Adaption in Natural and Artificial Systems*. MIT Press, Cambridge, MA, 1975.

[43] J. H. Holland. *Adaption in Natural and Artificial Systems*. University of Michigan Press, Ann Arbor, MI, 1975.

[44] S. Homberger and H. Gehring. Two Evolutionary Metaheuristics for the Vehicle Routing Problem with Time Windows. *INFOR*, 37:297–318, 1999.

[45] T. Hürlimann. *Mathematical Modeling and Optimization, An Essay for the Design of Computer-Based Modeling Tools*, volume 31 of *Applied Optimization*. Kluwer Academic Publishers, Dordrecht, The Netherlands, 1999.

[46] T. Ibaraki, S. Imahori, M. Kubo, T. Masuda, T. Uno, and M. Yagiura. Effective Local Search Algorithms for the Vehicle Routing Problem with General Time Window Constraints. *Transportation Science*, 2004, to appear.

[47] ILOG Optimization Suite. ILOG, Inc., Incline Village, Nevada, 2000. http://www.ilog.com/products/optimization.

[48] S. Irnich and D. Villeneuve. The Shortest Path Problem with k-Cycle Elimination ($k \geq 3$): Improving a Branch-and-Price Algorithm for the VRPTW. Technical Report G-2003-55, GERAD, Montreal, Canada, 2003.

[49] D. P. Jacobs, J. C. Peck, and J. S. Davis. A Simple Heuristic for Maximizing Service of Carousel Storage. *Computers and Operations Research*, 27:1351–1356, 2000.

[50] J. Kallrath. *Modeling Languages in Mathematical Optimization*. Kluwer Academic Publisher, Dordrecht, The Netherlands, 2004.

[51] G. Kartnig. *Wege zur Leistungssteigerung von Umlaufregallagern des Typs Rotary-Rack*. Habilitation, Technische Universität Graz, Fördertechnik, May 1997.

[52] S. Kirkpatrick, C. D. Gelatt, and M. P. Vecchi. Optimization by Simulated Annealing. *Science*, 220:671–680, 1983.

[53] W. K. Klein-Haneveld and M. H. van der Vlerk. Stochastic Integer Programming: General Models and Algorithms. *Annals of Operational Research*, 85:39–57, 1999.

[54] A. K. Klinger. *Spielzeitberechnung und Lagerdimensionierung des Typs Rotary-Rack*. Phd thesis, Technische Universität Graz, Institut für Allgemeine Machinenlehre und Fördertechnik, Abteilung für Fördertechnik und Machinenzeichnen, January 1994.

[55] G. Laporte, M. Desrochers, and Y. Nobert. Two Exact Algorithms for the Distance-Constrained Vehicle Routing Problem. *Networks*, 14:161–172, 1984.

[56] H. F. Lee. Perfomance Analysis for Automated Storage and Retrieval Systems. *IIE Transactions*, 29:15–28, 1997.

[57] H. F. Lee and S. K. Schaefer. Sequencing Methods for Automated Storage and Retrieval Systems with Dedicated Storage. *Computers and Industrial Engineering*, 32:351–362, 1997.

[58] S. Lin. Computer Solutions of the Traveling Salesman Problem. *Bell System Technical Journal*, 44, 1965.

[59] N. Metropolis, A. W. Rosenbluth, M. N. Rosenbluth, A. H. Teller, and E. Teller. Equation of State Calculations by Fast Computing Machines. *Chemical Physic*, 21:1087–1092, 1953.

[60] A. Mingozzi, L. Bianco, and S. Ricciardelli. Dynamic Programming Strategies for the Traveling Salesman Problem with Time Window and Precedence Constraints. *Operations Research*, 45:365–377, 1997.

[61] G. L. Nemhauser and L. A. Wolsey. *Integer and Combinatorial Optimization*. Wiley, New York, USA, 1988.

[62] I. H. Osman. Metastrategy Simulated Annealing and Tabu Search Algorithms for the Vehicle Routing Problem. *Annals of Operations Research*, 41, n.1-4:421–451, 1993.

[63] M. Padberg and G. Rinaldi. A Branch-and-Cut Algorithm for the Resolution of Large-Scale Symmetric Traveling Salesman Problems. *SIAM Review*, 33:60–100, 1991.

[64] C. H. Papadimitriou and K. Steiglitz. *Combinatorial Optimization: Algorithms and Complexity*. DOVER, Mineola, USA, 1998.

[65] J.-Y. Potvin and S. Bengio. The Vehicle Routing Problem with Time Windows-Part II: Genetic Search. *INFORMS Journal on Computing*, 8:165–172, 1996.

[66] J.-Y. Potvin, T. Kervahut, B. Garcia, and J.-M. Rousseau. The Vehicle Routing Problem with Time Windows-Part 1: Tabu Search. *INFORMS Journal on Computing*, 8:158–164, 1996.

[67] J.-Y. Potvin and J.-M. Rousseau. A Parallel Routing Building Algorithm for the Vehicle Routing and Scheduling Problems with Time Windows. *EJOR*, 66:331–340, 1993.

[68] J.-Y. Potvin and J.-M. Rousseau. An Exchange Heuristic for Routing Problems with Time Windows. *Journal of Operational Research Society*, 46:1433–1446, 1995.

[69] T. K. Ralphs. *Parallel Branch and Cut for the Vehicle Routing*. PhD thesis, Cornell University, Ithaca, NY, 1995.

[70] J. Rethmann and E. Wanke. Storage Controlled Pile-Up System, Theoretical Foundations. *European Journal of Operational Research*, 103:515–530, 1997.

[71] G. Righini. Approximation Algorithms for the Vehicle Routing Problem with Pick-up and Delivery. Technical Report 33, Note del Polo - Ricerca, July 2000.

[72] K. H. Rosen. *Handbook of Discrete and Combinatorial Mathematics*. CRC Press, Boca Raton, USA, 2000.

[73] ROTASTORE. psb. GmbH, Pirmasens, Germany, http://www.psb-gmbh.de/scripts/en/index.php.

[74] M. W. P. Savelsbergh. Local Search in Routing Problems with Time Windows. *Annals of Operations Research*, 4:285–305, 1985.

[75] M. W. P. Savelsbergh. An Efficient Implementation of Local Search Algorithms for Constrained Routing Problems. *European Journal of Operational Research*, 47:75–85, 1990.

[76] L. Schrage. Optimization Modeling with LINGO, 2004. LINDO Systems.

[77] R. Schultz. On Structure and Stability in Stochastic Programs with Random Technology Matrix and Complete Integer Recourse. *Mathematical Programming*, 70:73–89, 1995.

[78] R. Schultz. Stochastic Programming with Integer Variables. *Mathematical Programming Ser. B*, 97:285–309, 2003.

[79] F. Semet and E. Taillard. Solving Real-Life Vehicle Routing Problems Efficiently Using Tabu Search. *Annals of Operations Research*, v.41 n.1-4:469–488, 1993.

[80] M. M. Solomon. Algorithms for the Vehicle Routing and Scheduling Problems with Time Window Constraints. *Operations Research*, 35(2):254–265, 1987.

[81] M. M. Solomon, E. Baker, and J. Schaffer. Vehicle Routing and Scheduling Problems with Time Windows Constraints. In B. L. Golden and A. A. Assad, editors, *Vehicle Routing: Methods and Studies*, pages 85–106. North-Holland, Amsterdam, 1988.

[82] M. M. Solomon and J. Desrosiers. Time Window Constrained Routing and Scheduling Problems. *Transportation Science*, 22:1–13, 1988.

[83] H. Stadtler. Linear and Mixed Integer Programming. In H. Stadtler and C. Kilger, editors, *Supply Chain Management and Advanced Planning*, pages 335–344. Springer, Berlin, Deutschland, 2000.

[84] E. Taillaird, P. Badeau, M. Gendreau, F. Guertin, and J.-Y. Potvin. A Tabu Search Heuristic for the Vehicle Routing Problem with Soft Time Windows. *Transportation Science*, 31:170–186, 1997.

[85] P. M. Thompson and H. N. Psarafatis. Cyclic Transfer Algorithms for Multi-Vehicle Routing and Scheduling Problems. *Transportation Science*, 22:1–13, 1993.

[86] P. Toth and D. Vigo. *The Vehicle Routing Problem*. SIAM, Philadelphia, USA, 1st edition, 2002.

[87] D. Vanderbilt and S. G. Louie. A Monte Carlo Simulated Annealing Approach to Optimization over Continuous Variables. *Journal of Computational Physics*, 56:259–271, 1984.

[88] H.-J. Zimmermann. An Application-Oriented View of Modeling Uncertainty. *European Journal of Operations Research*, 122:190–198, 2000.

About the Author

Julia Kallrath was born in January 19, 1976 in Omsk, Russia. In 1992 she started her studies at the mathematical faculty at Omsk State University. Under the supervision of Prof. Dr. A. A. Kolokolov she finished her master's thesis on the modeling the social infrastructure of Omsk's region. In 2001 she started as a Ph.D. student at the Fraunhofer Institute for Industrial Mathematics (ITWM, Kaiserslautern, Germany) under supervision of Prof. Dr. S. Nickel. Her thesis was focused on online optimization problems and the solution of two real-life projects.

Index